Magnetic Relaxation in Iron Phthalocyanine Thin Films

finite-sized magnetic chains

BY
THOMAS GREDIG
PAUL D. EKSTRAND

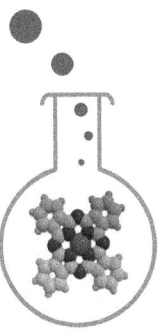

2005
GredigLab Publishing

Title image: temporal remanence curves measured at different temperatures for iron phthalocyanine / silicon thin film

Email: mail@thomasgredig.com
Web: www.thomasgredig.com

Gredig, Thomas
Ekstrand, Paul D.
    Magnetic Relaxation in Iron Phthalocyanine Thin Films.
    First edition.
    ISBN-13: 978-1532898341

Preface

At the University of Hamburg in 1922, Wilhelm Lenz gave then 22-year old Ernst Ising a problem to explore properties of ferromagnetism from a theoretical perspective. Subsequently, Ising concluded that the spin-spin correlation length of an infinitely long and highly anisotropic spin chain is vanishingly small at all temperatures above absolute zero. This important result has been re-derived in a modern description on page 71. It meant that in one dimension a two-state spin chain cannot undergo a phase transition from paramagnet to ferromagnet. This problem became a paradigm of theoretical physics and the derivation is recounted in numerous textbooks, homework problems, and examples. After a paper by Peierls, spins with two states became widely known as Ising spins. In 1944, Onsager formally solved the two-dimensional Ising system, which has a famous transition point at finite temperature from a paramagnetic system to a ferromagnetic system. The exact analytical solutions of an infinitely long Ising chain and a 2D Ising plane provide deep insight into magnetism, cooperative behavior, phase transitions, and more. The Onsager model has been confirmed by experiment in a variety of systems. Yet, the one-dimensional system of Ising remains a largely un-probed field in experimental condensed matter physics.

In 2001, the first single chain magnets with random lengths have been reported, and in 2004, adding defects and extra impurities gave a way to probe the chain length. Here, we present a system where crystals are limited in size during growth to probe the length dependence of magnetically insulated iron chains. Using a metallo-organic molecule to form chains made of iron ions, we can probe finite-sized systems in one dimension. The chain length can be adjusted experimentally in this system from about 100 to 300 ions by choosing appropriate deposition parameters.

The magnetic dynamics of low-dimensional iron ion chains have been studied with regards to the tunable finite-sized chain length using iron phthalocyanine thin films. The deposition temperature varies the diffusion length during thin film growth by limiting the average crystal size in the range from below 40 nm to 110 nm. Using a method common for single chain magnets, the magnetic relaxation time for each chain length is determined from temporal remanence data and fit to a stretched exponential form in the temperature range below 5 K, the onset for magnetic hysteresis. A temperature-independent master curve is generated by scaling the remanence by its relaxation time to fit the energy barrier for spin reversal, and the single spin relaxation time. The energy barrier of 95 K is found to be independent of the chain length. In contrast, the single spin relaxation time increases with longer chains from under 1 ps to 800 ps. We show that thin films provide both the nano-architecture to control magnetic relaxation and the testbed to study finite-size effects in low-dimensional magnetic systems.

This work is based on the M.S. thesis of Paul D. Ekstrand that was approved by the Physics and Astronomy Department at the California State University Long Beach on December 2015. The project was planned and guided by Dr. Thomas Gredig as a result of investigating the magnetic relaxation observed in first order reversal curves measured by graduate student Mathew Werber. The initial work started with a project for undergraduate student Daniel Javier. The samples in this work have been prepared by graduate students Tarun Sharma and Matthew Byrne. The VSM measurements were performed by graduate students Paul D. Ekstrand, Matthew Byrne, undergraduate student Daniel Javier, and Dr. Thomas Gredig. The analysis of the data was performed by Dr. Thomas Gredig using R. The iron phthalocyanine source material was purified by undergraduate student Brian Cacha. This work was partially supported by NSF grant DMR-0847552. Insightful conversations with Alessandro Vindigni are gratefully acknowledged.

Thomas Gredig
March 2017
Huntington Beach, CA

Contents

1 Introduction

Magnetic materials are important components in most modern day electronic equipment. Magnetic storage, speakers, and energy distribution all rely heavily on ferromagnetic materials. Conventional ferromagnetic materials exhibit spontaneous magnetization, also called remnant field, in the absence of an external applied magnetic filed. Such materials play a key role in motors, speakers, and data storage, but also have several limitations. Traditional ferromagnetic materials are inflexible, relatively heavy, and are manufactured at high processing temperatures. One alternative for lighter flexible materials with ferromagnetic or ferromagnetic-like properties are organic-based molecules. This class of materials may also have intriguing optical properties. In particular, metallo-organic thin films may be able to reduce or eliminate several of these disadvantages and provide new interesting applications reserved for flexible magnets.[57, 65] These emerging magnetic materials provide pathways towards magnets with novel optical properties.[61]

One of the unique and interesting properties of organic materials is their customizability. They are chemically tunable, which allows for many adjustments to be made to the individual molecules.[7] Due to this feature it is possible to tune the properties of the molecules to obtain desired behaviors. Many organic materials can be made to have the magnetic or conductive properties of metals.[97] In addition to this, organic magnets can be made using solution chemistry as opposed to the more difficult and energy intensive high temperature metallurgical methods of traditional magnets.[58] Due to the non-covalent bonding of organic molecules in a thin film, different structural phases provide a means to control important characteristics within one material.

Given the cooperative nature of magnetism and the linked strong interactions between neighboring atoms, obtaining strong magnetic properties in new organic-based materials is a challenge. Therefore, an interesting property of organic materials is the formulation of single molecule magnets, which occurs when a material is made up of individually magnetic molecules.[16] At the moment, single molecular

1

magnets remain mostly interesting at temperatures well below room temperature. Another class of materials, called single chain magnets, have prospects for ferromagnetic-like properties at higher temperatures up to room temperature.[59, 60] Recently, magnetic hysteresis has also been reported in iron phthalocyanine thin films.[40, 3, 27]

This introduction begins by outlining some of the basic theoretical understanding of classical magnetism. After this, information about metallo-organic molecules is added and specifically iron phthalocyanine thin films are discussed.

1.1 Magnetism

Magnetism has many facets. Fundamentally, all atoms formed of a nucleus and electrons have a magnetic moment. This magnetic moment is produced by moving and spinning charges. Due to the much higher mass of the proton and neutron, the moment from the nucleus is three orders of magnitudes smaller and will not be of concern here. Rather, the magnetic moment of the electron is the sole focus. From the atomic point of view, one generally defines a common unit, which is the Bohr magneton μ_B, defined for the electron's magnetic moment based on the mass of the electron m_e,

$$\mu_B = \frac{e\hbar}{2m_e} \tag{1.1}$$

This quantity can be derived from the semi-classical picture of an electron orbiting the nucleus and a quantized angular momentum. The small electron current I can be expressed in terms of the speed v, the orbit's radius r, and the charge e of the electron as $I = ev/(2\pi r)$. The speed must be given by the angular momentum. Classically it is written as $v = L/mr$, but in quantum mechanics, the angular momentum is quantized. Henceforth, the smallest angular momentum is $L = \hbar$, which produces Eq. 1.1 considering the magnetic moment as the product of area and current.

The gyromagnetic factor g is the ratio of the magnetic moment μ in units of Bohr magnetons and the angular momentum in units of $\hbar$.[83] For simple orbital motion, we would expect $g = 1$, and for the electron's spin contribution $g = -2$. In general, the g factor takes on more complicated values. In a classical picture, one would argue that

g represents the correction due to a non-uniform charge distribution inside a spherical structure. The total angular momentum $\vec{J} = \vec{L} + \vec{S}$ of an atom is expressed as the sum of the angular moment $\vec{L}$ and the electron spin $\vec{S}$. The moment of an atom is given by $\vec{\mu} = g\mu_B \vec{J}$. By including both the spin and angular component of the electron, the z-component of the moment can be expressed as[1]

$$\mu_z = (gm_s - m_l)\mu_B. \tag{1.2}$$

Here, m_s and m_l are quantum numbers. The magnetic moment interacts with a magnetic field $\vec{B}$. The Hamiltonian for an atom in a magnetic field is

$$\mathcal{H} = -g_J\mu_B \vec{J} \cdot \vec{B}. \tag{1.3}$$

This energy component is termed Zeeman energy. Atoms in an external field will tend toward a state of lowest energy. When the magnetic moment of the atom is aligned with the field, $\vec{J} \cdot \vec{B}$ is positive, making the Hamiltonian negative. This shows that the Hamiltonian is minimized when the spin and field are aligned parallel.[76]

In statistical mechanics, the Hamiltonian is plugged into the partition function in order to obtain the bulk magnetization $\vec{M}$. The partition function is easily solved under the restrictive condition that spins are non-interactive; i.e. there is no interaction between neighboring spins. Using this assumption, the result is a paramagnetic material with net magnetic moment that depends on temperature T and the magnetic field B. The result can be expressed in terms of the unitless quantity x, which depends on the ratio of the Zeeman energy and thermal energy $x \equiv (g\mu_B JB)/(k_B T)$. The net magnetization depends on the total number of atoms N and is derived to be

$$M(x) = Ng\mu_B J \cdot B_J(x). \tag{1.4}$$

We will introduce a new function called the Brillouin function $B_J(x)$, which has the following form

$$B_J(x) \equiv \frac{2J+1}{2J} \coth\left(\frac{2J+1}{2J}x\right) - \frac{1}{2J}\coth\left(\frac{x}{2J}\right). \tag{1.5}$$

In the classical limit of $J \to \infty$, the first term becomes $\coth(x)$, and the second term can be expanded with $\coth(x) \sim 1/x$. So, the Brillouin

---

[1]g can be negative

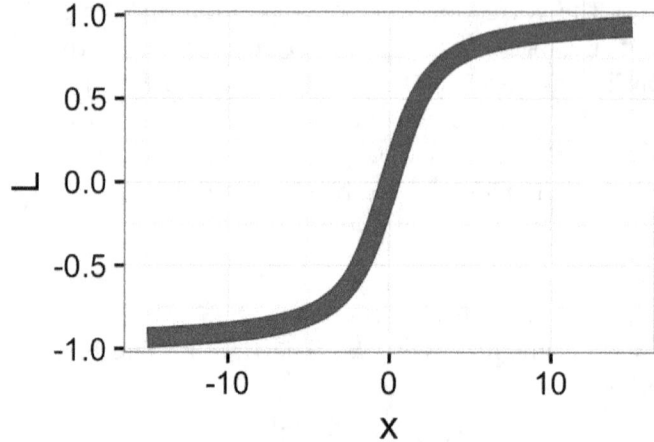

Figure 1.1: In the classical limit, the net paramagnetic moment of a material can be expressed with the Landé function. The x-axis shows a unitless quantity which represents the ratio of the Zeeman energy and thermal energy, namely $x \equiv (g\mu_B J B)/(k_B T)$.

function becomes the more easily conceptualized Landé function $L(x)$, which is

$$L(x) \equiv \lim_{J \to \infty} B_J(x) = \coth(x) - \frac{1}{x}. \tag{1.6}$$

The graph in Fig. 1.1 shows the general form of Eq. 1.6. It can be observed that this curve is linear for small values of x. In particular, you notice that $L(x) \sim x/3$ for $x < 1$. The temperature-dependence of the magnetization is derived from Eq. 1.4 in the classical limit with $J \to \infty$ as

$$M_{J \to \infty}(x) \approx N(g\mu_B)^2 \frac{B}{k_B T} \simeq C\frac{B}{T}, \tag{1.7}$$

which is known as the Curie law. It is applicable in the regime where T is large compared to the applied magnetic field B and the assumption of independent spins is valid. The constant C depends on the size of the sample (N), and contains the magnetic moment. In the quantum mechanical regime, the Curie constant is derived for small values of J and is written as

$$C = Ng^2 J(J+1)\frac{\mu_B^2}{3k_B}. \tag{1.8}$$

At low temperatures, in particular, interactions between spins become important. This overlap is captured in the concept of the exchange integral. Common materials, such as Fe, Co, and Ni, undergo a phase transition at a critical temperature T_c from paramagnetic phase to a ferromagnetic phase. One manifestation of the ferromagnetic phase is that the material can maintain magnetization, even in the absence of an external magnetic field. This magnetization is called the remnant magnetization and its state depends on the history of how it was obtained.

In ferromagnetic materials, the exchange interactions maintain spin coherence locally even in the absence of an external magnetic field. The net result are magnetic domains with spin coherence, which vectorially add to a net magnetization depending on the previous magnetization history of the material. The Hamiltonian from Eq. 1.3 cannot justify this outcome and needs to be augmented to properly predict ferromagnetic behavior. The exchange interaction between spins is a result of the Pauli exclusion principle; i.e. two fermions cannot be in the same quantum state. Assume that the electrons of two neighboring atoms a and b are brought close enough for the wave functions to overlap considerably. At this point, quantum mechanics does not allow us to distinguish the electrons any longer and the Pauli Principle leads to the conclusion that the 6-dimensional wave function is antisymmetric. Consequently the electrostatic potential becomes spin-dependent. Two electrons with parallel spin cannot come close together in space or else they would occupy the same orbital state. Electrons with anti-parallel spins will not have this requirement and the electric potential energy is slightly higher. To account for this effect, a term referred to as the Heisenberg Hamiltonian is added,

$$\mathcal{H} = -g\mu_B \vec{S}\vec{H} - 2J_{ex}\sum_{i,j}\vec{S}_i\vec{S}_j. \tag{1.9}$$

Here J_{ex} is a parameter which describes the strength of the exchange interaction. The J_{ex} coefficient is based on the spatial integration of the electron orbital overlap. Consider two atoms, a and b, sharing two electrons, k and l. The integral J_{ex} is just the integral for evaluating the energy of a Hamiltonian, $\int \psi^* \mathcal{H} \psi \, dr$. J_{ex} is therefore defined as

$$J_{ex} \equiv \int \int \psi_{a,k}^* \psi_{b,l}^* \left(\frac{e^2}{4\pi\epsilon_0 r_{k,l}^2} \right) \psi_{a,l} \psi_{b,k} d\vec{r}_k d\vec{r}_l. \qquad (1.10)$$

For very weakly interacting atoms, $J_{ex} \approx 0$, so that there is no significant exchange interaction and the material is considered paramagnetic. When $J_{ex} > 0$, the material is ferromagnetic and when $J_{ex} < 0$, the material is antiferromagnetic. For ferromagnetic materials, the global minimum of energy, in a classical sense at least, is obtained, when all neighboring spins are aligned simultaneously with each other and the applied external field. Due to the small spatial extent of wave functions, especially in insulators, only nearest-neighbor interactions need to be included in Eq. 1.9.

For magnetic materials, the susceptibility χ of a material can be written

$$\chi \equiv \frac{\partial M}{\partial B} = \frac{C}{T - \Theta} \qquad (1.11)$$

This is an extension of Curie's law (Eq. 1.7) and known as the Curie-Weiss law. The Weiss constant Θ is related to the critical temperature T_c, at which the phase transition from paramagnetic state to ferromagnetic state occurs in zero magnetic field. The coupling strength is estimated from the Weiss constant, if the spin S and number of nearest neighbors z are known

$$\Theta = 2z J_{ex} \frac{S(S+1)}{3k_B}. \qquad (1.12)$$

Ising showed that a highly anisotropic one-dimensional spin model cannot have a ferromagnetic transition at any temperature above absolute zero (see page 71). In 1944, Onsager showed that the same model in two dimensions leads to a critical temperature $k_B T_c \simeq 2.27 J_{ex}$ with the appearance of a non-zero magnetization at finite temperatures. This result is further described in a recent review.[6] The 3D Ising model is intractable.

1.2 Metallo Phthalocyanine

Organics are carbon-based compounds studied primarily for their optical and electrical properties. While the molecules of organic crystals are generally held together through weak van der Waals forces, the atoms inside the molecule are bonded through strong covalent bonds to form thermally and chemically stable complexes. Organic complexes can be classified in three ways: small-molecule organics, polymers, and biological organics.[33]

Organic-based semiconductors exhibit a wide range of properties such as photoconductivity, electroluminescence, fluorescence and phosphorescence, metallic conductivity, photovoltaic effects, magnetism and superconductivity, which is why they represent a very important group of materials from a basic science and application point of view. The main advantages over inorganic semiconductors are determined by the ease and affordable processing and integrating them with a wide variety of plentiful materials such as glass and plastic. These advantages give them the potential to be used in many applications such as large-area displays, solid-state lighting, radio frequency identification tags, solar cells, sensors or computing.[81]

Small-molecule based organic semiconductors have been intensively studied and integrated into a wide range of electronic devices. Due to their more highly ordered crystalline and polycrystalline structures, they are better suited than polymers for fundamental studies. Within this class of material, a menagerie of molecules with different structures and properties have been synthesized. Pentacene ($C_{22}H_{14}$) is a molecular semiconductor that has been grown in highly ordered thin films, and is investigated as a main candidate for organic thin film transistors.[78] Perylene ($C_{20}H_{12}$) is a polycyclic aromatic hydrocarbon which displays blue fluorescence and is used as a blue-emitting dopant material in pure and substituted organic light emitting diodes (OLED). Because an LED is merely a photoconductor operated in reverse, perylene is also being investigated for applications in solar cells. Small molecules can also contain metals. Often, a metallic ion is connected via ligands to form a stable metallo-organic molecule. A good example is the family of metallo-phthalocyanine molecules. Phthalocyanines have traditionally been used as organic dyes, but have found wide use as active materials in OLEDs, photovoltaic cells, and gas sensors.[96]

Phthalocyanines (Pcs) are the archetype of planar small molecules. It is a metal complex with the chemical structure of M-$C_{32}N_8H_{16}$, where M is either a metal ion or the H_2 complex.[91] Phthalocyanine has a history of being studied in the form of powder and thin films due to

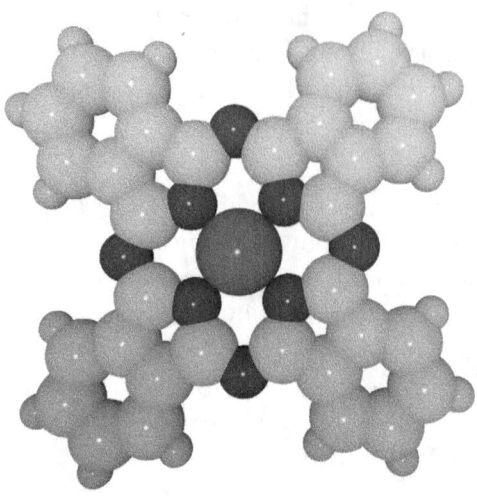

Figure 1.2: Metallo phthalocyanines are the archetype of planar small molecules and have the chemical structure of M $C_{32}N_8H_{16}$, where M is a metal ion at the center.

its good thermal and chemical stability.[18] The chemical structure is shown in Fig. 1.2. This molecule exhibits many intriguing physical and chemical properties that make it important as a class of materials in industrial, medical, and computational applications. Pcs are used commercially as blue and green pigments, where its color arises due to intense absorption between 600 nm and 700 nm.[18] Pcs are also commonly used as catalysts in manufacturing processes.[18] Moreover, Pcs show interesting photophysical properties, redox properties, and both photo- and dark-semiconductivity, which render them particularly interesting for use as photoconductors in the charge generation layer of drums in photocopiers and printers.[41] Another application of Pcs is found in optical data storage systems.[1] These properties have also led to research for use of Pcs in photovoltaic cells[95] and fuel cells.[51]

Metallo-phthalocyanines (MPcs) have been studied for many decades and as a result many of their structural and electronic properties are well understood.[52] As such, they serve as a model for the class of small planar organic molecules.[17] MPcs are generally low-mobility p-type semiconductors and are associated with unique structural and optical properties.

The first MPc synthesized was copper phthalocyanine (CuPc) in

1927.[24] Shortly after this discovery, the structural and chemical properties of CuPc were extensively studied.[54, 55] These studies were followed by measurements of the transport properties for CuPc.[28] The elements most commonly used as the central metal ion are copper, cobalt, nickel, iron, zinc, manganese, and lead. In the case of FePc, the Fe^{2+} cation has four bonds with nitrogen. It is a synthetic macrocyclic tetradentate ligand with a square planar coordination. MPcs are symmetric and have strong interactions between the metal centers, which leads to self-assembly. Molecules of MPcs form columnar structures in high aspect ratio. Molecule-molecule interactions are greatest when the planar molecules stack face to face as opposed to edge on. This affinity in growth causes molecules to stack with the central metal atoms forming a chain as illustrated in Fig. 1.3. The chain is parallel to the substrate on non-interacting substrate, such as silicon or sapphire, and perpendicular to the substrate on interacting substrates, such as gold. These chains are considered to be one dimensional with highly anisotropic magnetic, optical, and electric properties.[58] The most common type of crystalline structure formed by MPcs in powder is monoclinic. The molecules in the crystalline structure will pack themselves in a herringbone arrangement, that is they will form an angle with respect to the b-axis made by the central metal atoms.[94] The herringbone arrangement of the molecules is used to define the two dominant phases of MPcs in powder form. The first phase, designated α, has a small tilt angle of 26.5° with respect to the b-axis. The second phase, designated β, has a greater tilt of 44.8°.[2] The phase of the Pc crystal can be controlled during the crystallization process. The α-phase is meta-stable and can be transformed to the stable β-phase using thermal energy. X-ray diffraction is used to distinguish the two common phases. There are many other phases distinguished, including a thin film phase.[2]

One of the reasons MPcs are interesting to study is their high level of customizability. This has lead to the idea that the properties of MPcs might be tailored to specific applications. The central metal ion is one way to change the properties of MPcs. For example, CuPc has higher mobility and electrical reactance than FePc. Phthalocyanines containing transition metals like Co, Fe, Mn, and Cr have open d-shells with more complex electronic structures compared to the closed-shell MPcs like MgPc or ZnPc.[53] Exchanging the peripheral hydrogen

Figure 1.3: Typical growth of metallo-phthalocyanine on non-interacting or insulating substrates on the left. The red circles represent the central iron atom, while the lines show the ligand of the FePc molecule. This configuration is referred to as "standing". On an interacting surface, such as atomically flat gold, the chains grow vertically, known as "lying" configuration, shown on the right.

ligands with fluorine will change the MPcs from a p-type semiconductor to an n-type.[9, 96] Moreover, the electrical conductivity of MPcs can be increased by orders of magnitude by doping them with iodine.[72] Oxygen and water (H_2O) are also dopants, which are readily absorbed by MPcs.[62, 63]

The crystalline structure of MPcs can also be tailored.[71] On substrates where the molecule-substrate interaction is greater than molecule-molecule interaction, MPc chains will grow lying flat to the substrate.[4] This is contrasted with insulating substrates like silicon, where MPcs self-assemble in a standing configuration with the b-axis parallel to the substrate, see Fig. 1.3. One can then propose complex surfaces, such as gold nano-particle covered silicon surfaces, which induce specific growth patterns.

When MPc has a transition metal as its central atom, it is called a d-block MPc. Iron, for instance, has its outermost electrons in the 3d shell. Understanding the magnetic behavior of d-block MPcs requires consideration of the molecular properties, which depend on the electronic configuration of d-shells, and the intermolecular magnetic interactions, which depend on the crystalline structure.[45] The electrons in the d-shell exhibit ligand field splitting.

A ligand is an ion or molecule which binds to a central metal atom and creates a coordination complex. This binding requires the donation of an electron pair. In FePc, four pyrroles represent the immediate ligands. Ligands are usually viewed as electron donors and the metals as electron acceptors. When the two are bound, the ligand provides both electrons to the bond instead of the ligand and metal each contributing one. The binding of the metal to the ligand results in a new set of orbitals from which the metal can be identified with a new lowest unoccupied molecular orbits (LUMO) and highest occupied molecular orbits (HOMO) and a certain ordering of the d-

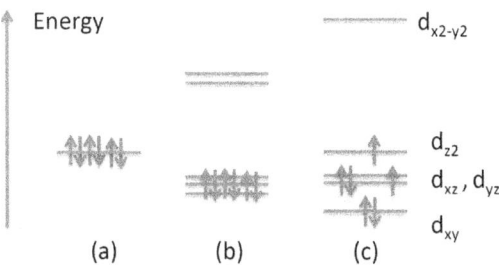

Figure 1.4: Schematic energy levels of the d-orbitals in (a) free ion, (b) octahedral coordination, and (c) tetragonal configuration. In the octahedral configuration, the 5 d-orbitals are degenerate and the bottom state is referred to as t_{2g} and the top state as e_g. This state is S=0. For the tetragonal configuration, S = 1, there are 4 distinct energy levels.

orbitals.[45] In the octahedral configuration, the otherwise degenerate d-orbitals split into three lower, called d_{xz}, d_{yz}, d_{xy}, and two upper energy levels, denoted as d_{z^2} and $d_{x^2-y^2}$. The spacing of these energy levels is referred to as octahedral field splitting energy Δ. A schematic representation of the energy levels is given in Fig. 1.4. First-principles study have been performed to compute the projected density of states for FePc on Au(111)[43] and free-standing FePc.[90] Wang et al. show that the electron distribution shown in Fig. 1.4 (c) is overly simplified. Computations of the magneto-crystalline anisotropy energy indicate in-plane easy axes for FePc and CoPc. In contrast, the anisotropy is strongest in MnPc, with the predicted anisotropy axis perpendicular to the molecular plane.[90] Experimentally, the in-plane easy axis in FePc is confirmed for thin films grown on Au via XMCD.[4] In the case of FePc/Si thin films, with the standing molecular configuration, this picture is complicated, since there is a herringbone-structure present. In this case, the zig-zag structure of the in-plane anisotropy cancels the component perpendicular to the b-axis, such that there is a net moment along the b-axis. Additionally in the standing configuration, the b-axes of many FePc crystals on Si are distributed randomly, so that some are aligned with the applied magnetic field, while others only have a projected value.

FePc is one of several MPc complexes that are known to show magnetic hysteresis and ferromagnetic-like[2] behavior.[45] Historically, the

---

[2]non-equilibrium superparamagnetism

magnetic behavior of FePc powder in the α-phase was first observed in 2001.[29] The magnetization versus applied field measurements for FePc powder show hysteresis, if measured at temperatures below 5 K. In the α-phase, the magnetic moments of Fe(II) ions are strongly coupled along the ferromagnetic chains via super-exchange. Evangelisti et al. speculated that this would be the signature of a canted, soft molecular ferromagnet below 10 K.[30] Given the magnetic nature of α-phase FePc powder, susceptibility data was taken from 50 K to 300 K (in the high temperature regime). This data was fitted to the Curie Weiss law, Eq. 1.11 with the assumption of negligible spin-spin interaction. In this framework, Evangelisti et al. extracted $\Theta \simeq 40$ K,[30] which corresponds to a nearest neighbor intra-chain coupling strength of $J/k_B \sim 15$ K, using the effective spin $S = 1$ and $z = 2$. Based on the assumption of a low-dimensional magnet, an Ising model was fit to the low-temperature powder susceptibility data for $\alpha-$FePc. In this model, the coupling strength is $J/k_B \sim 25$ K - slightly higher - and anisotropy energy is given by $D/k_B \sim 53$ K. The positive sign of D indicates that the magneto-crystalline anisotropy has in-plane easy axes in agreement with the earlier mentioned XMCD experiment. The Goodenough-Kanamori rules suggest weak ferromagnetic coupling that is mediated through super-exchange using 90° coupling with nitrogen.[46] The 90° coupling is due to the herringbone structure. In contradiction to the in-plane anisotropy, it was argued that α-phase FePc powder can be classified as an Ising-type chain system, analogous to $CsCoCl_3$, $(NH_4)_2MnF_5$, and $FeCl_2Py_2$.[3] As alluded, the zig-zag structure of the molecular planes cancel the component, which is perpendicular to the plane. Using the temperature-dependent Green function method, Oguchi[68] finds a relation between the intra-chain coupling strength J and the weaker inter-chain coupling strength J'. For FePc powder, an upper limit of 10^{-2} K is determined. Therefore, the ratio of $J/J' > 1000$ supports a categorization into a low-dimensional magnetic system.

In contrast to the α-phase, the structural properties of FePc in the stable β-phase have been intensively investigated since 1935.[54] In 1968, it was determined that the β-phase shows no magnetic order unlike the α phase mentioned above.

In ligand field theory, a simple molecular orbital study of metal ions with D_{4h} symmetry shows that the d orbitals have energies according to

---

[3]Pyridine (Py) has chemical formula C_5H_5N.

$e_g < a_{1g} < b_{2g} < b_{1g}$.[50] However, generally accepted assignment has been based on the study by Dale et al.,[22] in which they deduced $(b_{2g})^2$, $(e_g)^3$, $(a_{1g})^1$, configuration and $S = 1$ ground state from magnetic susceptibility measurement in the temperature range 1.25 K to 300 K, see Fig. 1.4. The data in the high temperature region, 100 K to 300 K, obeyed the Curie-Weiss law. They noted that often oxygen molecules get trapped and provide paramagnetic background, also ferric oxide (Fe_2O_3) can contribute to the background at low temperatures. The value of the effective magnetic moment per molecule was found to be 3.71 μ_B at room temperature. The gyromagnetic ratio g is anisotropic, since there is a large second-order spin-orbit coupling energy (~ 100 K). The computed values are $g_\parallel = 1.93$ and $g_\perp = 2.86$.

Unlike the powder form, thin films deposited from room temperature up to 230 °C show magnetic behavior similar to the α-phase in the powder form. The crystalline phases of FePc were determined in powder (i.e. bulk) samples. The precise structure in thin films is not clearly known.[2, 4] Because the chains of FePc grow parallel to the substrate, it is difficult to ascertain the exact geometry of stacking via x-ray diffraction. From structural measurements, we know that there is a small variation of the d-lattice spacing with the deposition temperature.[58] Also, we know that thin films deposited above 200 °C have grain size distributions with longer tails.[35] From atomic force microscopy, it appears that pinholes appear more prominently in those high-temperature thin films. Therefore, we argue that all data presented here is comparable with the α-phase in powder.

1.3 Growth of Iron Phthalocyanine

In high-vacuum deposition, epitaxial growth is determined by minimizing the Gibbs free energy as temperature and pressure are fixed in this environment. The chemical potential μ is representative of the work that needs to be done to add one molecule and is given by $\partial G/\partial n|_{p,T}$. The forces acting between the adsorbate and the substrate depend on the charge transfer, chemical bonding, electrostatic forces, and van der Waals interactions. Therefore, the growth of the first layer and subsequent layers may be different in particular if the substrate is an insulator.[58] Additionally, there are step-edge barriers which play a role in the growth.[34] This barrier of energy ΔE_{ES} is mathematically described by Ehrlich and Schwöbel and leads to the growth of mounds.[42, 26, 80] Experimentally, the jump rate ν is activated by this barrier and can be expressed as $\nu' = \nu_0 \exp\left[-\Delta E_{ES}/k_B T\right]$. The grain size d_g is related to the jump rate through a power law and depends on the deposition flux F, such that $d_g \sim (\nu'/F)^{1/5}$.[34, 85]

All MPc have a similar molecular structure with the only difference being the metal in its cavity. They can be considered isomorphic. The growth for different MPc is similar. The molecules of FePc stack by minimizing the energy when the metal-ions are close. Subsequently, iron chains are formed during crystallization. On Au(111) surfaces, the FePc lies flat, making the b-axis almost perpendicular to the substrate, such that the chains grow perpendicularly to the substrate surface. Ideally then, the thickness of the thin film corresponds to the iron chain length. Practically, this is difficult to achieve beyond a certain thickness, and evidence points towards limited thickness as measured through coercivity.[93]

The deposition temperature changes the diffusion length and changes the crystallite size during thin film deposition. Recently, Gentry et al.[35] showed quantitatively that the size of an FePc grain is proportional to the deposition temperature of the substrate. FePc grains are analyzed using an atomic force microscope (AFM). Images from the AFM show that grains are rod-like in shape. When samples are deposited on room temperature substrates, the grains are more circular-like. Raising the deposition temperature increases the length of the major axis faster than the minor axis. Considering only the largest 10% of grains (by area) it was shown that the major axis increases an average of 6 times, while the minor axis increases an average of 4 times. This occurred as the deposition temperature of the samples was brought from room temperature up to 260 °C. The relationship between the chain length along the major axis and the deposition temperature is phenomenologically exponential.[35]

There are also changes in the roughness of the MPc thin films as the deposition temperature is increased. The small mounds become more irregular, and above 200 °C pin holes are common in the structure. The distribution of grain sizes is fundamentally different, which encourages us to distinguish two regimes, namely the low temperature deposition (LTdep) for $T < 200$ °C and the high temperature deposition (HTdep) regime.

The grain size is also affected by the manner in which the grains grow. On a substrate made of a conducting material, the iron ions of FePc will tend to interact with the substrate surface. This will cause the major axis of the grains to grow perpendicularly to the surface of the substrate. In contrast, an insulating substrate will cause the FePc grains to grow parallel to its surface. The roughness of the substrate can also affect how the grains grow. Morphologically smooth gold will cause the major axis of the FePc grains grow perpendicularly to the substrate surface.

1.4 Low Dimensional Magnetic Systems

It is well established that in equilibrium, an infinite chain of Ising spins cannot have long-range interactions at any temperature above absolute zero.[44, 67, 8] For a derivation of the correlation length of the 1D Ising model, see page 71. Spin fluctuations due to the exchange energy will randomly remove any net magnetization as equilibrium is attained. For this reason one dimensional Ising chains have no hysteretic behavior to measure. In contrast, the 2D and 3D Ising analogues have a well-known ferromagnetic phase transition at finite temperatures.

Interestingly, there are two systems that have recently emerged that are low-dimensional, but still show hysteretic magnetization behavior and other ferromagnetic-like properties. In zero dimensions, there are single molecular magnets, independent molecules with a large spin and a strong Ising-like anisotropy. In one dimension, there are single chain magnets, long chains of metallic spins forming chains. The chains are generously spaced, so that intra-chain interactions are negligible.

Single molecular magnets (SMM) exhibit magnetism from a purely molecular origin.[19] Due to the collective behavior, long range magnetic ordering is not necessary for magnetism to be present. This contrasts with conventional bulk magnetic materials. The most well-known example is manganese-12-acetate, which has the chemical formula $[Mn_{12}O_{12}(OAc)_{16}(H_2O)_4]$.[12] Acetate is a salt that can be written as $(CH_3CO_2)^-$, water is added to stabilize the molecule.

Single chain magnets (SCM) exhibit slow relaxation of the magnetization and thus have magnetic properties comparable to those of SMMs before reaching full equilibrium. They provide, at low temperatures, a magnetic hysteretic behavior for a single polymeric chain.[19] The mechanism underlying the slow relaxation of SCMs can be studied with Glauber dynamics. Mathematically, the kinetic behavior of a one-dimensional Ising system is studied at finite temperatures. In SCM materials, the slow relaxation of the magnetization is not the signature of isolated anisotropic complexes as with SMMs, but rather arises from the magnetic interactions between anisotropic repeating units along a single chain.[19] SCMs are built with spin units presenting a strong uniaxial anisotropy expressed mathematically in the constant D. This means that SCMs have a preferred direction for the magnetic moment, known as the easy axis, which is energetically favorable and stable.[86] If a magnetic field saturates the moment of a SCM in the direction of the easy axis, then the remnant moment will be similar to the saturation moment. If a field is applied in any direction other than the easy axis, the moment is forced into a less stable state. The direction of least stability is the hard axis. As this is a state of higher energy than the easy axis, it will take a larger magnetic field to saturate the sample in this direction. When the applied magnetic field is removed the magnetic moment of the sample will move to a more stable (lower energy) state along the easy axis.[86]

The typical characteristic of a single chain magnet is generally measured as the product of susceptibility χ and temperature T, χT. From $\chi T \sim \exp\left[\Delta E / k_B T\right]$, the activation energy ΔE is determined and J_{ex} can be computed knowing the spin of the system.[92] The frequency dependent susceptibility data is analyzed for the unitless Mydosh parameter[13]

$$\phi = \frac{\Delta T / T}{\Delta \left(\ln f_p\right)},$$ (1.13)

where f_p is the peak frequency. The Mydosh parameter can distinguish between a single chain magnet and a spin glass, which also has a frequency dependent susceptibility. In FePc powder samples, it was measured that $\phi = 0.034$, which is about 6 times larger than a spin-glass, and one order of magnitude smaller than a superparamagnet. [30] Susceptibility measurements in powder samples are a common characterization method, however, in thin films, the substrates contribute significantly to the total signal and susceptibility measurements become very difficult or unfeasible to measure.

An open question remains whether FePc is a single chain magnet, or whether intra-chain interactions are strong enough to form a two- or three-dimensional ferromagnetic system. In the former case, the organic ligands of FePc keep chains magnetically insulated from each other when in a crystalline structure such as a thin film. This reduces the collective interaction of the spins, meaning that the magnetism of the FePc samples is consistent with SCM behavior. In support of this, the $J \gg J'$, but also the fact that the asymptotic magnetization M_∞ tends to go to zero. Another question is whether the anisotropy of FePc is strong enough to be considered Ising-like.

While ac susceptibility provides one approach to measure time-dependent properties, magnetic properties can also be probed over time. Previously, it has been reported that FePc thin films show magnetic hysteresis, when measured at temperatures below 5 K.

A typical hysteresis loop is formed by sweeping the magnetic field to a maximum positive applied field, back to the same field in the opposite direction and returning to the positive field, see Fig. 1.5. This process takes a particular time and relaxation times are usually much faster than measurement speeds. The applied field H-intercepts are known as positive and negative coercivity. In a ferromagnetic material, this is the required strength of the applied field to reduce the magnetization to zero. The M-intercept value defines the remnant magnetization of the material. The remnant magnetization is the amount of magnetization left when the applied field is reduced to zero. The area inside a hysteresis loop is the energy dissipated due to internal forces. Each of these characteristics is important in the study of magnetic materials and their properties.

Since the hysteresis loop plots magnetization as a function of applied field, there is no time dependency represented in the graph. For samples with slow relaxation, these curves cannot capture the full information. For FePc thin film samples, the remnant magnetization and coercivity are field sweep rate dependent; i.e. the rate at which the magnetic field is swept co-determines the values for coercivity and remnant magnetization.

In the following the remnant magnetization is measured over time. The spontaneous magnetization relaxes in the absence of an external magnetic field. This characterizes the dynamic response to changes in the field.[5] The remnant magnetization relaxes at different time scales. In other materials, relaxation times of a few seconds for small

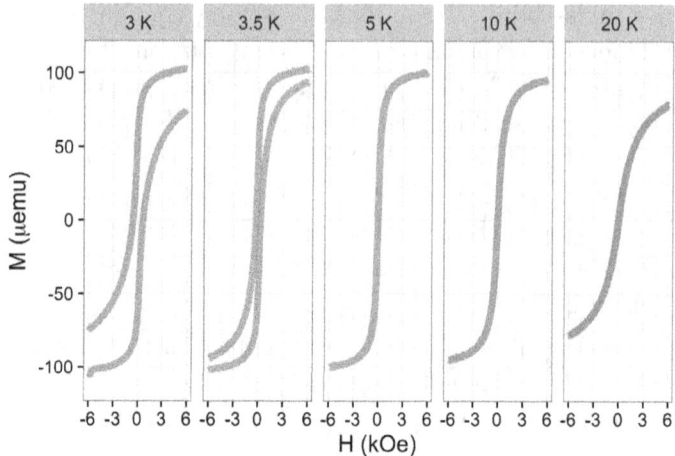

Figure 1.5: Hysteresis loops at 5 different measurement temperatures for FePc thin film of nominal thickness 150 nm deposited at 150 °C. For measurement temperatures below 5 K, magnetic hysteresis is observed, and along with hysteresis, there is a non-zero coercive field as well as a remnant magnetic field. Magnetic fields of 3 T are applied to saturated the film at all temperatures. The diamagnetic substrate background has been removed.

magnetic particles at room temperature[5] have been observed. On the other end of the scale, relaxation times up to several hours were measured for Fe clusters below 1 K.[79] Slowly relaxing single molecular magnets (SMMs) are an intense area of research as they would provide some of the densest forms of information storage, with a bit stored by a single molecule. Slow relaxation was first shown in single chain magnets (SCMs) in 2001. It can be argued that industrial applications are easier to realize with SCMs than SMMs.[79]

1.5 Glauber Dynamics

In the pioneering work of Nobel laureate Roy Glauber a mathematical framework was constructed in 1963 to explain the statistical nature of magnetic relaxation in systems with Ising-like spins.[36] At finite temperatures, the individual spins interact with the heat bath and can fluctuate randomly. In the Glauber-Ising model, spins depend on the values of neighboring spins in addition to the heat bath. Considering a chain of N spins under Markov process, the time-dependence of the magnetization is determined, and the frequency-dependent susceptibility is analyzed in the weak-field limit. In the absence of a magnetic field, the Hamiltonian for a linear Ising chain is written as

$$\mathscr{H} = -J\sum_{i=1}^{N}\sigma_i\sigma_{i+1} \tag{1.14}$$

The model is probably best understood with a simple example for a single spin, or $N = 1$. This approach also exemplifies the mathematical process. In this single-spin system, the Ising state $\sigma_1 = \pm 1$ is not allowed to flip randomly, but rather at a pre-determined rate α, so that the differential equation governing the state of the spin at any time is explicitly stated

$$\frac{d}{dt}p(\sigma,t) = -\frac{\alpha}{2}p(\sigma,t) + \frac{\alpha}{2}p(-\sigma,t). \tag{1.15}$$

The boundary condition that $p(1,t) + p(-1,t) = 1$ has to be met at all times. The pair of equations can be reduced to one function $q(t)$, which if chosen to be

$$q(t) \equiv p(1,t) - p(-1,t) = \sum_{\sigma=\pm 1}\sigma p(\sigma,t) \tag{1.16}$$

resembles the expectation value of the magnetization; i.e. $q(t) = <\sigma(t)>$. Taking the time derivative and using Eq. 1.15, we get four terms that add up to $dq/dt = -\alpha q$, and therefore has a solution

$$q(t) = q(0)e^{-\alpha t} \tag{1.17}$$

Using the insight of $q(t)$, it then follows that

$$p(\sigma,t) = \frac{1 + \sigma q_0 e^{-\alpha t}}{2}. \tag{1.18}$$

For a multi-spin system ($N > 1$), we need to introduce a second parameter. In addition to the transition switching rate α, namely, the neighbor interaction parameter γ. The pre-factor $\alpha/2$ is now written in the form of the transition probability $w_i(\sigma_i)$, which can take on three values

$$w_i(\sigma_i) = \begin{cases} \frac{\alpha}{2}(1-\gamma), & \sigma_{i\pm1} \text{ are parallel} \\ \frac{\alpha}{2}, & \text{if } \sigma_{i-1} = -\sigma_{i+1} \\ \frac{\alpha}{2}(1+\gamma). & \sigma_{i\pm1} \text{ are antiparallel} \end{cases} \qquad (1.19)$$

Using the simple Ising Hamiltonian in the absence of a magnetic field, see Eq. 1.14, we can solve for γ and find[36]

$$\gamma = \tanh\left(\frac{2J}{k_B T}\right). \qquad (1.20)$$

This model is overly simple and real systems are not expected to have such simplicity built in. In order to steer theoretical models towards usable solutions, experimental data must be collected.

To summarize the Glauber-Ising model, there are 5 parameters to keep in mind: the (static) neighbor exchange interaction J, the anisotropy energy D, the number of spins N, the flipping rate α, and the (dynamical) neighbor interaction parameter γ.

This shows that over time the average of the spins will tend to either zero, or an offset determined by boundary conditions. In our experiment, the relaxation data is fit to a stretched exponential function.

1.6 Stretched Exponential Function

The relaxation behavior of complex, slowly relaxing and strongly interacting materials often shows non-Debye like behavior due to a distribution of relaxation times.[69] Several empirical mathematical descriptions have been proposed, including power-law forms, such as Cole-Cole form $f(\nu) \sim 1/[1 + (2\pi i\nu\tau)^{1-\alpha}]$ and Davidson-Cole form $f(\nu) \sim 1/(1 + 2\pi i\nu\tau)^{\delta}$, Havriliak-Negami form $f(\nu) \sim 1/[1 + (2\pi i\nu\tau)^{\alpha}]^{\beta}$, and the stretched exponential form.[25] The stretched exponential function was first documented by Kohlrausch in 1847.[73, 74] In mathematics, this function became known as the complementary cumulative Weibull distribution. Waloddi Weibull (Swedish) developed the statistical theory of particle size distributions from fracture. This has led to the statistical distribution called Weibull distribution,

$$f(x) = \frac{k}{\lambda}\left(\frac{x}{\lambda}\right)^{k-1} \exp\left[-\left(\frac{x}{\lambda}\right)^{k}\right], \qquad (1.21)$$

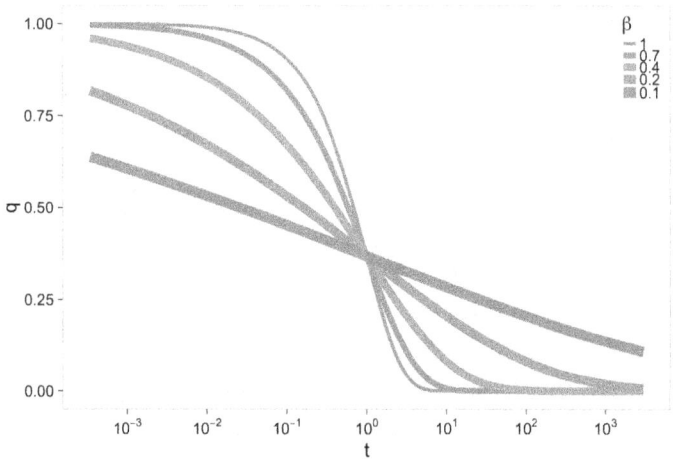

Figure 1.6: The stretched exponential function $q(t)$ is graphed for several values of β ranging from 0.1 to 1 for fixed $\tau = 1$ on a semi-log plot.

for $x > 0$. The two adjustable parameters are known as the shape parameter k and the scaling parameter λ. For $k < 1$, failures become more rare over time; i.e. most defects occur in the beginning. Despite the naming of the function, earlier work had been done by Rosin and Rammler on coal powder particle distributions.[77] Recently, it was shown that using a physical derivation of the Weibull distribution based on a fragmentation process also shows - under certain circumstances - the similarity with the log-normal distribution.[10]

Much before this development, Kohlrausch found that the decay of residual charge in a Leyden jar (an early capacitor) could be described with a stretched exponential function, which is a cumulative Weibull function,

$$q(t) = q_0 e^{-(t/\tau)^\beta}. \tag{1.22}$$

Fig. 1.6 shows the effect of the parameter β on the shape of the function for a fixed relaxation time.

This empirical relaxation function, later also known as the Kohlrausch - Williams - Watts form, enjoys near universal success in describing relaxation behaviors with a particular relaxation time τ. The stretch exponent β must take a value $0 < \beta < 1$. In the case of $\beta = 1$, the relaxation is Debye-like or "conventional". A β value greater than one would turn the function into a compressed exponential, the most famous example of which is the Gaussian function ($\beta = 2$).

The stretched exponential function is not the result from a fundamental microscopic theory yielding particular attributions to the stretched exponential β. So, we cannot attribute a microscopic understanding to this parameter yet. Essentially, the stretched exponential function is regarded as a convenient experimental and phenomenological tool without a deep fundamental significance.[14] A brief summary of the incomplete understanding for the significance for the stretching parameter β is provided now.

For most of the existence of the stretched exponential function, the stretch exponent was seen as an adjustable parameter with no intrinsic significance. Although the use of stretched exponentials are hard to justify physically, it can be shown that broad distributions of relaxation time τ will lead to smaller values of the stretched exponent.[84]

Probably the simplest way to generate the stretched exponential form is to start with the Debye-like relaxation, $q(t) = q_0 \exp(-t/\tau)$. Assuming that there are additive contributions with different relaxation times, each having a different weight $w(\tau)$, we can then write:

$$q(t) = \int_0^\infty w(\tau) \exp^{-t/\tau} d\tau \qquad (1.23)$$

Palmer et al. have discussed several plausible distributions for the function of $w(\tau)$.[69] Some interesting cases can be solved analytically and yield as a result the stretched exponential function.

In 2011 Naumis et al. reported that for homogeneous glasses β bifurcates into two rational "magic" values, $\beta_{sr} = 3/5$ and $\beta_K = 3/7$ under some assumptions.[66] According to them, the values of β can be shown to fall into one of these two values when glassy samples are made with high microscopic homogeneity. This can, in fact, be used to measure the quality of glasses as the numerical distance from these two values corresponds to inhomogeneities in the measured glasses.[66] When short range forces dominate the material, β takes a value of $3/5$. When long range Coulomb forces are present in the material, β takes on a value of $3/7$. This behavior has been found in a wide range of materials including: polymers, electrolytes and alcohols, spin glasses,[64] water, fused salts, and heme proteins to mentioned a few.

Finally, we want to comment on the broad application of the stretched exponential function. It has been used to characterize oil field size distributions, temperature variations at Vostok, stock market price variations, and citations of physicists.[47]

1.7 Scaling Law

Many systems can be described effectively by a single dynamical equation. For spin glasses with an extremely wide distribution of relaxation times, the entire data for the imaginary part of the complex dielectric constant $\epsilon''(\omega, T)$ can be scaled onto a single curve with the assumption that the relaxation times are related to the distribution of activation energies.[21] The measurable sample conductance $G(\omega) \sim \omega \epsilon''(\omega, T)$ can be measured for several temperatures over a fixed range of frequencies. A scaling function $R(u) = 0.5(1 + \tanh(Au))$ is introduced with A as a constant. In this case $u = B - (T - T_0) \ln(\tau/\tau_0)$, where B is a constant cutoff energy, so that the imaginary part of the dielectric constant can be expressed as

$$\epsilon'' = h(T)R(u)\left(1 - x\ln(\omega\tau_0)\right) \qquad (1.24)$$

The implication is that the relaxation time distribution is independent of temperature. At the same time, and this is the advantage of this exercise, temperature can be used to extend the temporal range of the function. In other words, many short measurements over several temperatures correspond to the knowledge of a very long measurement in time.

Traditionally, the relaxation times in single chain magnets are obtained from the ac susceptibility measurements. The frequency-dependent peak of the out-of-phase susceptibility component shifts with measurement temperature. As the temperature is increased, the frequency peak shifts towards higher frequencies with lower amplitudes. A Cole-Cole plot of the susceptibility shows a half-circle that can be modeled with a polydisperse model.[49] For each temperature, the relaxation time can be extracted. In thin films, however, this method is unsuitable as the substrate makes significant contributions to the signal.

Hence, in thin films, a master curve of the remanent magnetization $M(t, T)$ versus a scaled time t/τ is generated from many temperature measurements. This is similar to previous work.[31, 49] This is based on the assumption that the energy barrier is independent of the measured temperature within a range. As an example, Ferbinteanu et al. extract relaxation times τ over 8 decades using the collapsed data along with ac and dc measurements.

In the following analysis two methods for analysis are examined. In the first analysis technique the remanent magnetization $M_0(t)$ of FePc thin films is fit with a fixed value of $\beta = 0.43$ ($\sim 3/7$) using the stretched exponential fitting model, with the following dependence

$$M_0(t) = M_\infty + \Delta M e^{-(t/\tau)^\beta}. \tag{1.25}$$

The fits are made with only 3 independent parameters, in this case M_∞, ΔM, and τ. Here M_∞ corresponds to the asymptotic relaxation value, an offset, ΔM is the dynamic response, and τ is the relaxation time. The non-zero asymptotic relaxation value would correspond 3D interactions; i.e. cross-chain coupling. One dimensional behavior would require that $M_{\infty \ t\to\infty} = 0$ vanishes for very long measurement times.

From this analysis, we concluded that the asymptotic relaxation value is highly dependent on the measurement time, since the stretched exponent is very low. Even though good fits are achieved, the fitting parameters show strong time-dependence. It shows that the experimental time span is insufficient for the fits to converge.

Consequently, in the second analysis method, the temperature dependent curves $M_0(t, T)$ are scaled to form a master curve into $M(t')$, which extends over 7 decades in time. The master curve is modeled with a single temperature-independent curve

$$M_0(t) = \Delta M e^{-(t/\tau)^\beta}. \tag{1.26}$$

In this analysis, a single exponent β is used to describe all temperature curves simultaneously. It suggests that the asymptotic relaxation value vanishes for sufficiently long measurements. This second method provides a basis for careful analysis of the relaxation time τ on the chain length L or average number of spins N in a chain. The results support the notion that chains are indeed magnetically insulated and all long-range ordering dissipates over time at all temperatures above 2.5 K. The anisotropy energy D is independent of the number of spins N in the chain. In contrast, the single spin relaxation time strongly increases with chain length.

2 Thin Films

Samples for this experimental study were prepared using purified FePc powder and cleaned silicon substrates. The FePc powder is obtained from Sigma-Aldrich and purified in a custom-built thermal gradient purification system.[89] Then, the powder is loaded into a quartz crucible surrounded by a tungsten basket. The material is first outgased for several hours in high vacuum ($<10^{-5}$ mbar). Once the base pressure is around 10^{-6} mbar, the FePc crucible is slowly heated to near $350\,°C$. At this temperature, typical deposition rates of 2–6 nm/min are achieved in the Nano-Master Thermal Evaporator 3000 (NTE-3000) via sublimation. Sample thickness and substrate temperature are carefully monitored using a quartz crystal monitor and thermocouple. The thickness is then calibrated via x-ray diffraction. Next, samples are measured in a Quantum Design physical property measurement system (PPMS) with vibrating sample magnetometer (VSM) module. The sample series includes 8 different deposition temperatures to vary the grain size. The applied magnetic field is in the same direction as the substrate plane. A final sample is co-deposited onto a gold covered silicon substrate to modify the growth direction of the b-axis, see Fig. 1.3. In the latter configuration, the applied magnetic field is parallel to the substrate and therefore perpendicular to the chain axis.

2.1 Thin Film Deposition

The substrates for the samples are cut from silicon wafers. Using a diamond tipped scribe, wafers are cleaved into 4 mm by 10 mm rectangles. These rectangles are then cleaned using a process with three steps. First, the substrates are sonicated in warm acetone for 5 minutes to remove organic impurities. The substrates are then immediately placed in methanol and sonicated for another 5 minutes, in order to remove acetone residue from the substrate. When removed from the methanol, the substrates are dried with nitrogen gas. The methanol mostly evap-

Table 2.1: A list of all measured thin film samples and their deposition temperatures T_{dep}, substrates (S), surface areas A, and measurement dates. Samples have a thickness of around 154 nm and were either deposited onto silicon (Si) or gold/silicon (Au/Si) surfaces. All samples were prepared in 2010 and measured either 2.5 or 5 years after deposition.

Sample ID	Name	T_{dep} (°C)	S	A (mm^2)	Date Measured
MB100604Si1	S32	32	Si	33.98	5/30/2013
TS100913Si1	S80	80	Si	31.92	4/08/2013
TS101008Si2	S100	100	Si	30.45	11/28/2015
TS101202Si1	S140	140	Si	35.58	4/15/2013
TS100917Si1	S160	160	Si	35.28	11/25/2015
TS101218Si1	S180	180	Si	31.30	4/19/2013
TS100921Si1	S220	220	Si	33.62	6/06/2013
TS100928Si2	S230	230	Si	38.45	10/14/2013
TS101218Au	A180	180	Au/Si	29.40	4/22/2013

orates and does not remain on the substrate surface. The substrates are immediately placed on the substrate holder to be placed into the evaporation chamber. Inside the evaporation chamber, the substrates are heated to about 100 °C before deposition in order to remove water from the surface one day before deposition.

Next, FePc thin films are deposited onto the prepared substrates from purified FePc powder. The FePc powder is purchased from Sigma-Aldrich at $\sim$90% purity. The FePc is purified several times with a Lindberg / Blue M tube furnace following the thermal gradient purification method described elsewhere.[89] Three glass tubes are placed in a line inside the furnace with the center tube containing the unpurified FePc powder. A vacuum is generated with a conventional roughing pump. The tube furnace heats the inner tubes with a gradient temperature, so that the center has the peak temperature of about 450 °C. The FePc powder is sublimated inside and re-sublimated in the cooler zones. In this method, volatile materials will pass the tubes and deposit as white crystals outside the collecting glass tubes. Impurities with higher sublimation temperatures remain in the center glass tube. The purified FePc can then be extracted from the middle portion of the tubes. The process is repeated several times to achieve high-quality material. A sample list with their important attributes is provided in Tab. 2.1.

The thickness of sample S230 was measured with x-ray diffraction and determined to be 154 nm thick. All samples were prepared using a quartz crystal monitor to record the deposition rate; so all samples are expected to have the same thickness. The sample area A was measured using a Celestron digital USB microscope with a background reference grid. The digital image was analyzed with NIH ImageJ software that allows measuring irregular areas. A small portion of the substrate, where it was clipped down during deposition, is excluded. Therefore, the substrate area is always larger than the FePc surface area.

The sample ID in Tab. 2.1 contains the preparation date in the first column. All samples were prepared in 2010 using high-vacuum thermal evaporation. The relaxation behavior was measured either in 2.5 years later, or in the case of 2 samples 5 years after sample fabrication. Since the sweep rate for two samples, S100 and S160 was different compared to all other samples, these two samples were re-measured in 2015 at the same sweep rate of 106 Oe/s. It is expected that some aging of the sample occurred in this time interval and the total magnetic signal is diminished for those two samples. Recent work suggests that the magnetic properties change due to water and oxygen exposure most during the first year after deposition. After that period, FePc thin films reproduce magnetic characteristics.[23] Herein all samples will be referred to by their short name also listed in Tab. 2.1. The deposition temperature is varied from 32 °C to 230 °C varying the grain structure.[35] Most samples were deposited onto cleaned silicon substrate surfaces, but one sample was deposited onto Au to vary the molecular arrangement from standing to lying configuration. Samples S180 and A180 were co-deposited to ensure that the deposition parameters are identical. Measurements were also performed close to each other, which should make the comparison valuable.

The silicon substrates are clipped into the holder, which is placed at the top of the NTE-3000's vacuum chamber. The powder FePc is placed into a ceramic crucible that can be heated with a wolfram wire. To begin the deposition, the vacuum chamber's pressure is pumped out to just below 10^{-6} Torr (1.3×10^{-6} mbar). A bake-out is then performed to remove any water on surfaces. When the bake-out is complete, the substrate is heated to the desired deposition temperature. The crucible of FePc is heated until the FePc begins to sublimate generally at temperatures around 350 °C. The powder in the crucible is heated

Figure 2.1: Optical image of sample S220 showing the rectangular shape of a typical sample. Most of the sample surface is covered with FePc, however a small area where the clip held the sample in place during deposition is bare showing the silicon substrate. The ridges at the bottom are separated by 1 mm to provide quantitative length scale. The area is determined using the software tool NIH ImageJ.

such that the powder on top of the crucible is hottest. Once the FePc begins sublimating, the temperature is controlled to maintain a steady deposition of between 1 and 2 machine units (MU) per second. It is noted that 1 MU/s corresponds to about 0.45 Å/s.[93] Deposition rate and thickness is measured *in-situ* using a quartz crystal monitor.

2.2 X-ray Diffraction

The total sample thickness was monitored with a quartz crystal monitor. For one sample, the thickness is calibrated via x-ray diffraction (XRD). For the XRD measurement a larger sample was used to maximize the diffracted signal. The witness sample (deposited at the same time with other samples) has an area of 10×10 mm^2. A Rigaku SmartLab diffractometer is used in Bragg-Brentano configuration with a 10 mm slit and K$_\beta$ filter. The XRD diffraction data is shown for angles of 2θ from $4°$ to $9°$ for sample S230L. The main peak at $6.93°$ is accompanied by a secondary peak at $6.28°$, see Fig. 2.2. Using the Bragg relation, the thickness of the thin film is determined to be 154 nm. All samples are made to have the same thickness.

Secondly, we confirmed that the growth of the FePc molecules on

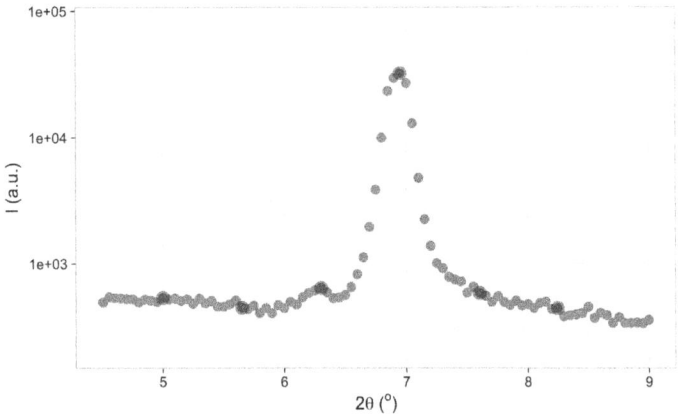

Figure 2.2: X-ray diffraction shows a strong peak at 6.93° corresponding to the main crystalline formation of FePc. Due to interference between the two beams off the FePc surface and the FePc/Si interface, there is a smaller peak at 6.28°, sometimes referred to as a Kiessig fringe. From the fringes, the precise thickness of 154 nm is determined. The larger blue circles mark the location where peaks are expected. Due to the roughness, only one fringe peak is clearly determined in this sample.

silicon is the same as on sapphire. Previously, the standing configuration had been studied in phthalocyanine thin films on sapphire substrates.[58] The α−FePc(200) peak is observed at $2\theta = 6.93°$. This corresponds to a d-lattice spacing of 12.7 Å in agreement with XRD measurements by Miller.[58] It has a Gaussian rms width of $\sigma = 0.123°$, which corresponds to a characteristic crystal size of about 28 nm according to the Debye-Scherrer relation. The silicon substrate peaks are also shown in Fig. 2.3.

2.3 Magnetic Measurements

The magnetic measurements were made using a Quantum Design physical properties measurement system (PPMS) with a vibrating sample magnetometer (VSM) module attached. In the PPMS, the applied field can go as high as 9 T, and the size of the field can be varied from a minimum rate of 10 Oe/s to a maximum sweep rate of 100 Oe/s. Samples are affixed to a quartz paddle with Kapton tape and placed in the sample chamber of the PPMS. The Kapton tape is used as it

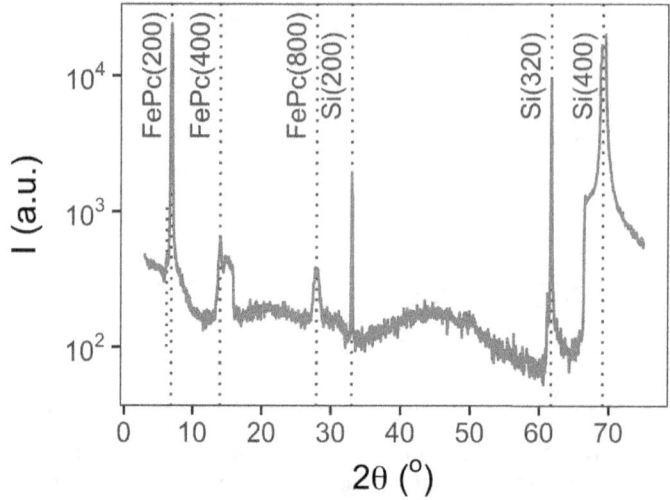

Figure 2.3: X-ray diffraction shows growth of FePc thin film on silicon substrate. In addition to the Si(200) at 32.99°, there are 3 peaks of FePc at 6.93°, 13.93°, and 27.98° corresponding to α-FePc(200), FePc(400), and FePc(800). The sample was deposited at 230 °C.

retains its adhesion even at low temperatures. Quartz is paramagnetic, but placing the sample in the center of the paddle will insure that any signal contributed by the quartz is cancelled out due to symmetry of the set up. The quartz paddle screws into the end of a meter long rod, which is then inserted into the PPMS. The rod attaches itself to the linear motor of the VSM, which moves the sample up and down at a set frequency of 40 Hz. The magnetic moment is picked up using a compact gradiometer pick-up coil configuration.

Care must be taken to not scratch or contaminate the thin film, keeping in mind its small sample volume of only $\sim 5 \times 10^{-6}$ cm^3. To this end, samples are handled with plastic tweezers to prevent the introduction of small magnetic metal particles. Latex gloves are worn to prevent contamination from oils of both the sample and other equipment. All sample preparations are done on Kimwipes to inhibit the introduction of dust particles on the sample. To facilitate securing the sample to the Quantum Design quartz paddle, a paddle holder is used. The sample is placed FePc side down. When securing the sample to the paddle, care must be taken to not slide the sample, as this would scratch and damage the sample. Enough Kapton tape is used to wrap around the sample exactly twice. This is done consistently so as to introduce the same background. This will better affix the sample to the holder while still remaining thin enough to not make contact with the sides of the sample chamber, which has an inner diameter of ~ 4.5 mm.

The VSM measures the magnetization of a sample by rapidly vibrating it vertically through a set of pick-up coils inside the PPMS. The changing magnetic field created by the rapidly moving sample induces a current in the coils. Using this induced current, the PPMS computes the magnetization of the sample using Faraday's law of induction and a model based on a point-like particle.

All magnetic relaxation measurements are made in the PPMS using the VSM module. To measure the time dependent magnetic relaxation we first saturate the sample in a 3 T magnetic field at a temperature of 100 K. This temperature is held for a period of 5 minutes. The sample is then cooled to 2.5 K, and the temperature is held there for 5 minutes to ensure temperature and magnetic field stability. The applied magnetic field is linearly removed as shown in Fig. 2.5. After the 300 s stabilitzation interval, the magnetic field is removed. Since the field cannot change by more than 100 Oe/s, it takes approximately 5 minutes to reduce the field from 3 T to 0 T. Once the field is at zero (or close to 0 Oe due to some remnant flux in the superconducting coils), the magnetic moment is measured and allowed to relax for 5000 s. After the relaxation measurement, the sample is warmed back to (at least) 100 K. The PPMS is then set to cool to the next temperature in the presence of a 3 T magnetic field to induce the same starting conditions. Relaxation was measured at 10 different temperatures between 2.5 K to 5.0 K. Data taken at 4.5 K and 5.0 K has no significant amount of remnant magnetization, and hysteresis loops do not show coercivity.

Data was taken using two different processes defined by the order in which the temperatures occur. The two processes, denoted A and B throughout, are described in Tab. 2.2. Changing the order of the measurement temperatures is thought not to change the results of the measurements. Process B was created due to issues involving the temperature stability of the sample chamber at the lowest temperatures. For some of the measurements, the PPMS was unable to cool down below 2.8 K. This caused the PPMS to continually heat and cool. By starting at 3.0 K, the PPMS was able to get useful data while the option

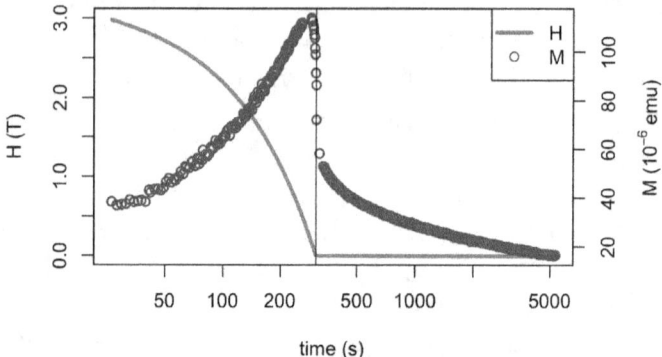

Figure 2.4: At $T = 3.2$K for sample S230, the magnetic field H is linearly reduced from 3 T to 0 Oe at ~ 100 Oe/s. Indicated with the vertical black line, at time $t \sim 310$ s, the remnant magnetization M is recorded for the next 5000s in zero applied field. Due to the silicon substrate, a diamagnetic background reduces the net measured magnetization M in the time span from 0s to 310s.

remained to take more data at a later date if the PPMS was unable to cool sufficiently.

Though the PPMS was programmed to reduce the magnetic field at a rate of 100 Oe/s, the measured rate was slightly faster at 106 Oe/s. This faster rate was reproducible for all measurements.

The PPMS has two built-in methods for stepping the time intervals during measurements. The first method is the standard "linear time", which takes data at regular intervals. The second method is "logarithmic time", which takes data at shorter intervals during the beginning of the measurement, and constantly increases the amount of time between each measurement. This means that more data is taken at the beginning of a measurement at the expense of data towards the end. For this experiment, the logarithmic time method was selected. Since the magnetization of the sample decreases rapidly at the start, it is important to take data fast initially. The fact that the parameters of the stretched exponential function converge quickly makes taking data at the beginning of the measurement more important. At the beginning of the measurements, the data points are still spaced, likely due to the fact that the magnetic field of the PPMS needed to be stabilized,

Table 2.2: Order of temperature measurements for two processes called A and B. The relaxation measurements are made at $H=0$ Oe for the listed temperatures.

	T (K)	
	A	B
1	2.5	3.0
2	2.6	3.2
3	2.8	3.4
4	3.0	3.6
5	3.2	3.8
6	3.4	4.0
7	3.6	2.5
8	3.8	2.6
9	4.0	2.8
10	4.5	4.5
11	5.0	5.0

providing only about 1 data point every 10 s, see Fig. 2.5.

2.4 Fit Robustness

The time span of 5000 s could be repeated without temperature stability issues of the PPMS. For longer measurement periods at temperatures below liquid helium temperature, the PPMS would heat up at times and thereby interrupt the measurements. Measurements with significant temperature fluctuations were removed from the analyzed data set.

The longest remnant magnetization measurement is 10 000 s and all measurements included at least 5000 s. In the following, we provide an analysis of what time scale should be used for a reliable fit. We apply the stretched exponential fitting method (page 24) to a particular portion of the data that has a time span Δt and determine the three fitting parameters. In addition, a second parameter called t_{min} is used to select the starting point. For each time span, three fits were performed, with the full data set. But, then also a fit for the data set starting at 200 s and 400 s, respectively. Since most of the drop of the magnetization occurs in the first few seconds, excluding this initial data shows how robust the fits actually are.

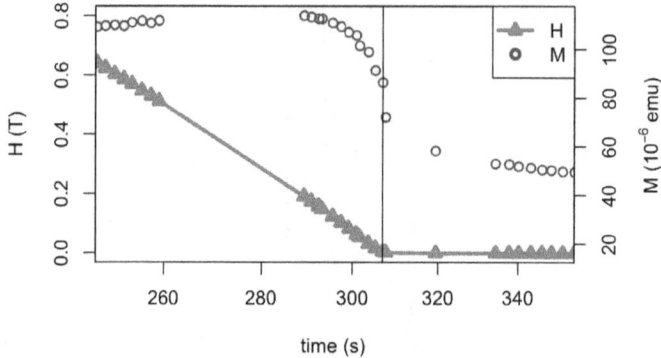

Figure 2.5: This figure is a partial view in of the data from 250 s to 350 s shown in Fig. 2.4. The magnetic field is reduced linearly. The start time is determined at the vertical black line at the point where the applied field is less than 5 Oe.

The asymptotic value M_∞ fluctuates for time spans under 2000 s, see Fig. 2.6. After that it slowly decreases with longer time spans indicating that even the stretched exponential fit does not capture the long-term trends to the magnetization. The removal of the initial 200 s or 400 s of data further drops the value of M_∞. For example, fitting the data from 0 s to 2000 s leaves $M_\infty = 40\,\mu\text{emu}/\text{cm}^2$; increasing the time span to 3400 s and 5000 s changes it to $37\,\mu\text{emu}/\text{cm}^2$ and $35\,\mu\text{emu}/\text{cm}^2$, respectively. We can expect that the converging value has not been reached yet. More fitting data points are listed in Tab. 2.3.

The dynamic reponse ΔM is a fitting parameter that measures the amplitude of the fitted function. It slowly increases with longer time spans and like M_∞ does not completely converge in the measured time span. The trend of ΔM is opposite to M_∞, yet the sum slightly decreases. The sum of M_∞ and ΔM represents the magnitude of the magnetization at $t = 0$ s. The actual magnitude, however, is $151\,\mu\text{emu}/\text{cm}^2$ and smaller than the sum of $163\,\mu\text{emu}/\text{cm}^2$ at 5000 s. Not surprisingly, if the initial data points are neglected, the fits yield smaller values for ΔM as shown for $t_{min} = 200$ s and 400 s in Fig. 2.7.

The third fitting parameter is the relaxation time τ. If the fitting time span Δt is increased beyond 2000 s, it increases slightly from 330 s

Table 2.3: Fitting parameters for different time spans Δt. As the time span is increased, the asymptotic magnetization M_∞ is decreasing and does not yet reach equilibrium at 5 ks. Similarly, the dynamic response ΔM increases with the fitting time span Δt. The relaxation time increases with longer fitting time span. The data is for sample S230 measured at 3.2 K.

Δt	M_∞	ΔM	τ
s	µemu	µemu	s
400	42	125	339
500	41	126	358
500	41	126	358
700	42	125	336
900	43	125	330
1100	42	125	333
1600	41	126	359
2300	39	127	386
3400	37	127	423
5000	35	128	466

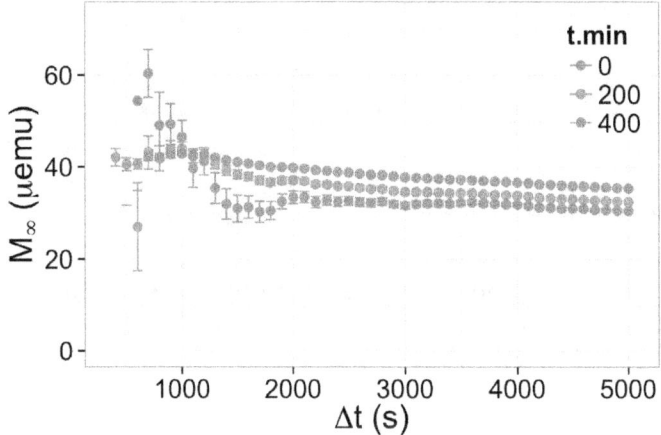

Figure 2.6: The asymptotic magnetization M_∞ is plotted versus the fitting time span Δt. Each data point represents the best fit for the data set that includes a subset of the complete data. The starting point is also varied t_{min}. Data was taken from sample S230 at 3.2 K.

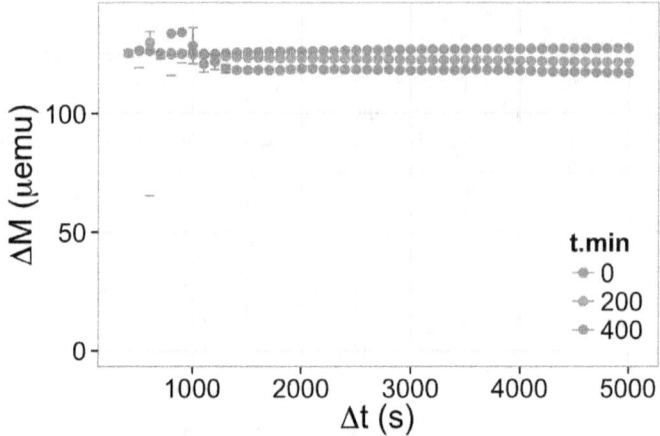

Figure 2.7: Fitting values of change in magnetization ΔM depend on the length of measurement time Δt as shown in the graph. As with asymptotic remnance, the dynamic response fluctuates strongly below 1500 s. Data was taken from sample S230 at 3.2 K.

at 900 s to 423 s and finally to 466 s at 3400 s and 5000 s fitting time spans, respectively. The increase is significant and represents a 29% increase in the relaxation time τ when compared to the 2000 s time span fit.

Although the fits visually look compelling (Fig. 3.2 for comparison), the 3 fitting parameters have some dependence on the measurement time. In particular, we find that longer measurements will increase the dynamic response, decrease the asymptotic magnetization, and strongly increase the relaxation times. Keeping these trends in mind, we will progress and fit measurements with at least 5000 s of data.

For all the measurements, the stretch exponent is fixed at $\beta = 0.43$. If the stretch exponent is left to fluctuate, it would add a forth adjustable parameter. It was found that it would vary from about 0.3 to 0.6 for all data sets; however, setting the stretch exponent to 0.43 ($\sim 3/7$) yields quite good fits.

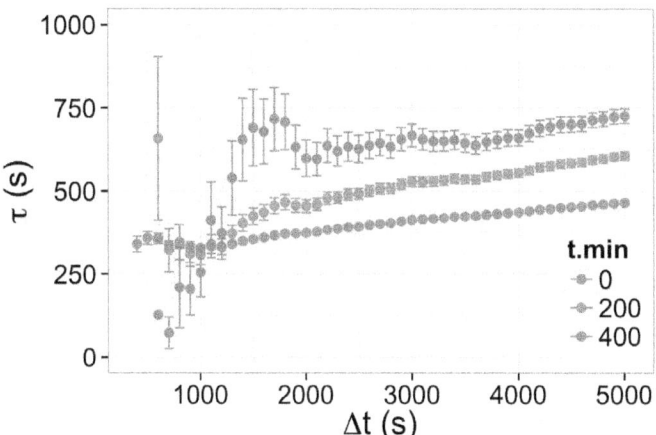

Figure 2.8: Fitting values of relaxation time τ depend on the length of measurement time Δt as shown in the graph. This parameter converges for $\Delta t > 1500$ s, showing that measurements of 5000 s provide sufficient accuracy. Data was taken from sample S230 at 3.2 K.

3 Relaxation Measurements

The magnetic properties of the powder form of iron phthalocyanine have been established by Evangelisti et al. around 2002.[30, 32] There, it is proposed that iron ions are strongly magnetically coupled into ferromagnetic chains coupled and (possibly) weak antiferromagnetic interchain coupling. Below 5 K, irreversible behavior was observed in powder samples that manifested itself in the form of magnetic hysteresis loops.

Later, thin films of iron phthalocyanine (FePc) were studied with the focus on magnetic hysteresis loops and it was shown that the chain length modifies the coercivity.[37, 40] The most important magnetic characteristics are generally inferred from ac susceptibility data. However, in thin films this approach bears complications as the thin film is mounted on a substrate. The substrate, although diamagnetic, can contribute significantly due to its large volume compared with the thin film. Furthermore, the thin film generally has a very small signal in comparison with the detection limit of the instruments. For the VSM option, the limit is a bit less than 10^{-6} emu at 1 s averaging.

In the following, the magnetic relaxation is studied as a function of temperature below 5 K, but also as a function of the iron chain length, which is inferred from the grain size. The grain size is tuned with the deposition temperature, in this case from about 39 nm to 190 nm, see page 52. The relaxation data of the magnetic remanence is fit to a stretched exponential function, which yields three fitting parameters, the asymptotic magnetization M_∞, the dynamic response ΔM, and the relaxation time τ using the first method (Equation 1.25). The temperature data is then collapsed into a master curve using the second method (Equation 1.26). From the temporal data spanning many orders of magnitude, a single stretching exponent β is derived that can be used to fit all data.

From the first analysis method, we find that samples with long iron chains tend to have higher values for the asymptotic magnetization. However, the asymptotic magnetization has not fully converged yet.

The dynamic response shows a peak at a characteristic temperature, which increases with the length of the iron chain. The magnitude of the dynamic response also increases greatly. The relaxation time near the transition temperature shows activated behavior with an energy barrier of roughly 3.2 meV ($\sim$ 38 K). The relaxation time increases with iron chain length. Since, the fitting parameters have not converged fully within the experimental time scale of 5000 s, a second analysis method is applied.

Rescaling the temperature-dependent relaxation data allows to generate relaxation data from saturation (obtained at low temperatures) to complete desaturation (measured at high temperatures). The stretched exponential function is fit over 7 decades of time. Using the resultant stretched exponent, a single fit parameter for each temperature is performed to obtain the relaxation time. We find that the relaxation time has the expected Arrhenius behavior for single chain magnets. The slope or energy barrier appears to be independent of the chain length. However, the single spin relaxation time strongly increases with chain length.

3.1 Remnant Magnetization Measurement

The temporal measurements of the remnant magnetization are recorded after saturating the sample in a magnetic field of 3 T. Hysteresis loops appear to be major loops for all samples measured as low as $T = 2.5$ K for this magnitude of the magnetic field. As illustrated in Fig. 2.4, the magnetic field is rapidly removed and then stopped when reaching $H = 0$ Oe. The net magnetization with the applied field of 3 T is relatively low, because the diamagnetic background of the silicon substrate reduces the net magnetic moment of the sample. Near $H = 200$ Oe applied field, the net magnetization reaches a peak and starts to decrease thereafter. While the magnetic field is reduced at a rate of 106 Oe/s continuously, the rate is decelerated quickly at the very end.

In the following 5000 s, the magnetization decrease in the absence of an applied magnetic field is recorded for a set of measurement temperatures. A typical curve measured at 3.2 K is shown in Fig. 3.1. The initial magnetic moment of 151×10^{-6} emu/cm^2 quickly drops to 134×10^{-6} emu/cm^2 after only 20 s. After the first 100 s, the net magnetic moment has dropped by 40 µemu/cm^2 or 26%. This sharp de-

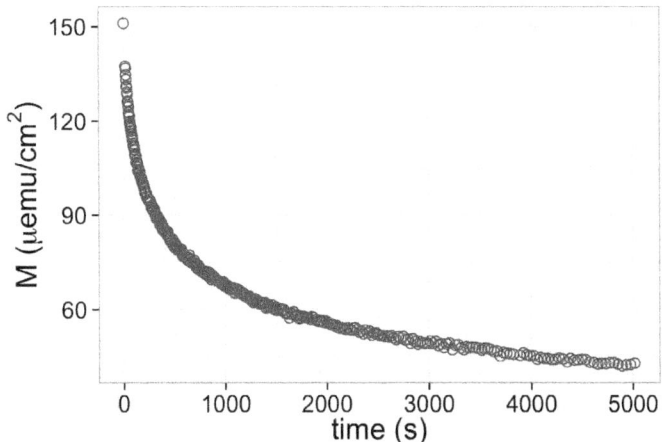

Figure 3.1: Typical relaxation behavior of the remnant magnetization is shown for a duration of 5000 s. Previously, the sample S230, FePc deposited at 230 °C, is saturated in a magnetic field of 3 T at $T = 3.2$ K. Here $t = 0$ s corresponds to the time, when the external magnetic field reaches 0 Oe.

crease can neither be modeled by a power law, nor a simple exponential function. After the first 1000 s, the magnetization reaches 68 µemu/cm^2 and the last recording at 5000 s shows 43 µemu/cm^2. Indeed, in the last 1000 s, the magnetic moment was reduced by less than 2%, but nevertheless equilibrium is not reached.

Fits to a simple exponential function and power law are poor. In fact a 3-parameter fit to a stretched exponential function, simple exponential function and power law shows that the best fit is with the stretched exponential function using the fixed stretching exponent $\beta = 3/7$. Optimization of the stretching exponent shows only small improvements. For the remainder of the discussion, a fixed stretched exponent $\beta = 3/7$ is used in all fits.

As a typical example of fitting the stretched exponential function to the data, Fig. 3.2 illustrates the close overlay of measured data with a single 3-parameter function. Sample S230, FePc/Si deposited at 230 °C, shows a dynamic response of $\Delta M = 128$ µemu/cm^2. The relaxation time τ is 467 s and the asymptotic magnetization M_∞ is 35 µemu/cm^2. The graph is plotted on a semi-log scale to emphasize the deviations of the fit function with the data at the very beginning of the data, $t < 30$ s and at the very end of the data, $t > 3000$ s.

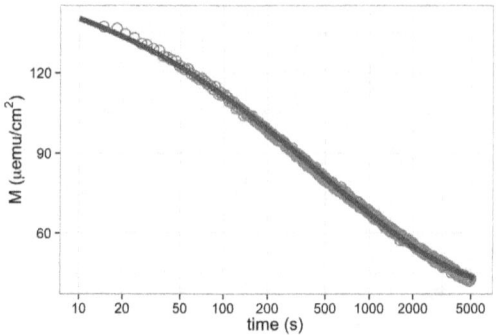

Figure 3.2: remnant magnetization measured at 3.2 K for sample S230. The experimental data is fit to a stretched exponential function with parameters $M_\infty = 3.53 \times 10^{-5}$ emu/cm^2, $\Delta M = 1.28 \times 10^{-4}$ emu/cm^2, and $\tau = 467$ s. The stretching exponent β is 3/7.

3.2 Stretched Exponential Fits of Remanence

Each FePc sample listed in Tab. 2.1 was measured at several temperatures between 2.5 K to 5 K. Measurements with significant temperature fluctuations were removed from the analysis. A set of measurements including 9 different temperatures for one sample is shown in Fig. 3.3. The magnetization behavior is reminiscent of those published for spin glasses.[15] The rapid decrease of the magnetization is best observed when the time is plotted on a logarithmic scale. In the final analysis, 5 to 9 different temperatures were analyzed for 8 different deposition temperatures. For all samples, the 4.5 K temperature measurement showed no measurable magnetic relaxation. For each temperature, the stretched exponential function (page 24) is fit using three independent parameters: the asymptotic magnetization M_∞, the dynamic response ΔM, and the relaxation time τ. The stretched exponent β was fixed at the fraction 3/7. A typical fit is shown in Fig. 3.2. While the fitted line closely follows the data in the middle part, there are noticeable deviations at long and short times.

It should be noted that there is a boundary condition:

$$M(t = 0 \ s) = M_\infty + \Delta M.$$

Experimentally, $M(t = 0 \ s)$ is known approximately and so a boundary condition is set. This makes the fit essentially a two-parameter

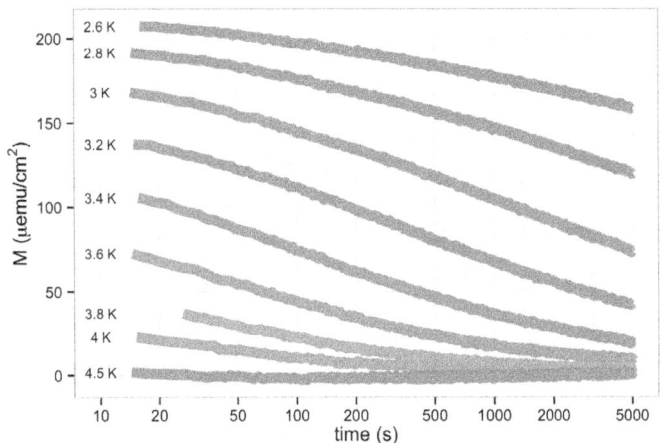

Figure 3.3: The time-dependence of the remnant magnetization for FePc/Si sample deposited at 230 °C show how the magnetization decays with time. For all measurements the magnetic field was reduced from saturation at 3 T to 0 T at a rate of 100 Oe/s. The measurement is started after the external magnetic field is removed and labeled as t=0 s.

fit. Even though the magnetic field is rapidly removed, it takes a finite amount of time. At the rate of 106 Oe/s, the time to drop the magnetic field from 3 T to zero, takes $\sim$ 300 s, a considerable amount of time, and as we shall see comparable to the relaxation time τ. We expect that the rate has an influence on the fitting parameters and shall only use data with a fixed magnetic field removal rate for the analysis.

The time for the start of the relaxation measurement; i.e. $t = 0$ s needs to be determined. For this analysis, once the applied magnetic field is less than 5 Oe, the next data point is defined as $t = 0$ s. The time step between measurements is generally around 3 s to 4 s, but during the magnetic field deceleration, time steps increase to around 10 s to 15 s. This limits the resolution of the starting time.

3.3 Asymptotic Magnetization

The asymptotic magnetization measures the amount of the remnant magnetization that should be available after waiting for a very long time. A non-zero value in equilibrium would also indicate interactions between chains. As mentioned in the experimental section (page 33),

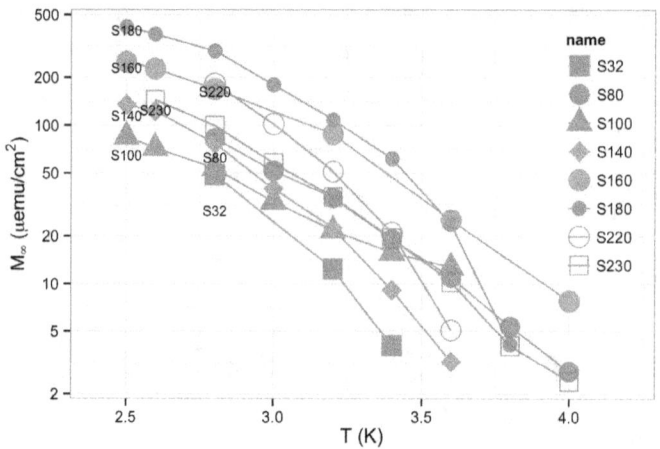

Figure 3.4: Values of asymptotic remnance are plotted against the measured temperature. Data can be fit with an exponential decay function.

the asymptotic magnetization does not converge, or reach equilibrium. It keeps decreasing without stabilization within the measurable time span. Hence, it should be interpreted as a upper limit, keeping in mind that M_∞ could be reduced further with longer measurements. A summary of the asymptotic value is shown in Fig. 3.4. On the semi-log plot, two features stand out. First, the longer iron chains lead to higher values of M_∞. Secondly, the value appears to vanish before reaching the 4.5 K threshold temperature, at which the coercivity vanishes. In fact, if the limit is taken where M_∞ drops below $5\,\mu\text{emu/cm}^2$, then successively longer chains have higher temperatures. For the shortest chain, the vanishing temperature is around 3.4 K, whereas it reaches above 4.0 K for the sample with the highest vanishing temperature, S160.

Tab. 3.1 summarizes the asymptotic magnetization values for all eight samples at several measured temperatures. Taking the values at 2.8 K, for example, one observes an increase in the asymptotic magnetization with chain size from sample S32 to sample S180. The last two samples deposited at temperatures above 200 °C show diminished behavior. Given that the sample roughness increases significantly for samples deposited above 200 °C and a significant change in the distribution of grain sizes was previously observed, it is conceivable that

Table 3.1: Asymptotic magnetization values M_∞ are obtained from a fit to remnant magnetization for 8 samples deposited at different temperatures. The asymptotic magnetization values strongly decrease with temperature and vanish near 4 K for all samples. These values were obtained from 5000 s long measurements. The results depend on this time. All magnetic fit values are given in units of 10^{-6} emu/cm^2.

	S32	S80	S100	S140	S160	S180	S220	S230
2.5			85.0	134.7	250.6	418.9		
2.6			72.0	123.0	227.4	377.1		144.4
2.8	48.8	82.6	53.7	76.9	169.0	294.4	182.7	99.5
3.0		51.8	32.9	39.8		179.7	101.9	57.9
3.2	12.4	34.9	21.7	22.4	88.2	108.4	51.3	35.3
3.4	4.1	19.1	15.7	9.1		61.8	20.9	19.3
3.6	-0.9	11.0	12.7	3.2	25.1	23.7	5.1	10.3
3.8	-4.3	5.4		-0.4		4.1	-4.5	4.0
4.0		2.8		-2.4	7.8	-6.8	-11.3	2.4

chains are interrupted despite the increase in overall crystal size.

3.4 Dynamic Response

The dynamic response ΔM strongly depends on temperature and relates to the amount of relaxation that occurs in a sample. The magnitude of the dynamic response increases for samples with longer iron chains, see Fig. 3.5. A peak is observed in ΔM for most samples and the temperature at which the peak occurs shifts to lower temperatures with shorter chain lengths. For example, the S180 sample has the highest value of $\Delta M = 325\,\mu\text{emu/cm}^2$ at 3.4 K, whereas sample S100 peaks near 3.0 K with a smaller value of $68\,\mu\text{emu/cm}^2$ as can be read in Tab. 3.2.

We use a parametrization of the resultant data to find out how the peaks shift with deposition temperature. Empirically, it can be seen that a Gaussian function fits the data well. Therefore, the dynamic response ΔM for each sample can be fit to the following function

$$\Delta M(T) = m_p e^{-\frac{(T-T_0)^2}{2\Delta T^2}}, \tag{3.1}$$

where m_p is the maximum peak height of the dynamic response for a given sample. The width of the peak is characterized by ΔT, which

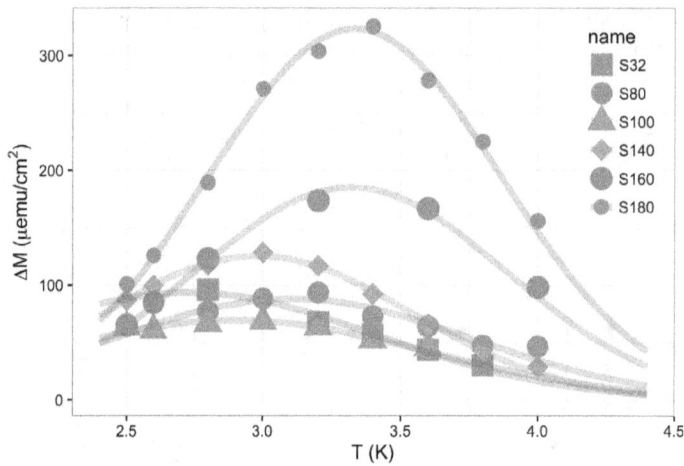

Figure 3.5: The dynamic response ΔM is plotted with respect to the measured temperature. The data has a clear peak which is estimated using a Gaussian fit.

Table 3.2: The magnetic dynamic response fitting values ΔM measure the maximum change in remnant magnetization during relaxation. For most samples a peak in temperature dependence is observed. All magnetic fit values are in units of 10^{-6} emu/cm^2.

T (K)	S32	S80	S100	S140	S160	S180	S220	S230
2.5			63	87	65	101		
2.6			61	99	85	126		71
2.8	96	77	66	118	123	189	174	102
3		88	68	129		271	225	128
3.2	68	93	64	117	174	304	236	128
3.4	57	73	52	92		325	214	115
3.6	44	64	46	65	167	278	166	90
3.8	30	47		43		225	116	58
4		46		29	98	156	78	38

Table 3.3: Fitting values from parametrization of the dynamic response $\Delta M(T)$ using Eq. 3.1. The peak center position T_0 shifts from 2.7 K to 3.3 K as the iron chain length increases in samples S32 to S180. The maximum peak height is characterized by the fitting parameter m_p.

sample	m_p $10^{-5}\,\mathrm{emu/cm}^2$	T_0 K	ΔT K	FWHM K
S32	9.34	2.73	0.684	1.61
S80	8.80	3.11	0.656	1.54
S100	6.93	2.90	0.704	1.65
S140	12.60	2.98	0.558	1.31
S160	18.50	3.32	0.564	1.33
S180	32.30	3.33	0.530	1.25
S220	23.50	3.19	0.520	1.22
S230	13.00	3.16	0.516	1.21

carries the units of temperature. The peak position is revealed by the fitting parameter T_0. The results from fitting Eq. 3.1 are listed in Tab. 3.3 and suggest that the peak temperature T_0 shifts from 2.7 K for the room temperature sample S32, all the way to 3.3 K for sample S180, which also has the largest peak amplitude and also the longest iron chains. From the Gaussian rms width, the he full width half maximum (FWHM) is computed. It decreases with chain length from 1.6 K for S32 to 1.3 K for S180. The amplitude of the dynamic response markedly increases with the iron chain length, indicating that the relaxation is much stronger in those samples.

3.5 Relaxation Time

The relaxation time τ is obtained from 5000 s long remnant magnetization curves. It was argued on page 34 that the relaxation time will increase with longer measurement time spans. In fact, Tab. 2.3 shows that the relaxation time can increase from 386 s at a measurement time span of 2300 s to 466 s at the full measurement time span of 5000 s. Therefore, a quantitative analysis is limited, yet we can analyze the set qualitatively. The results of the relaxation time τ are listed in Tab. 3.4 and graphed in Fig. 3.6.

The first point is that the relaxation times above 3.5 K become short,

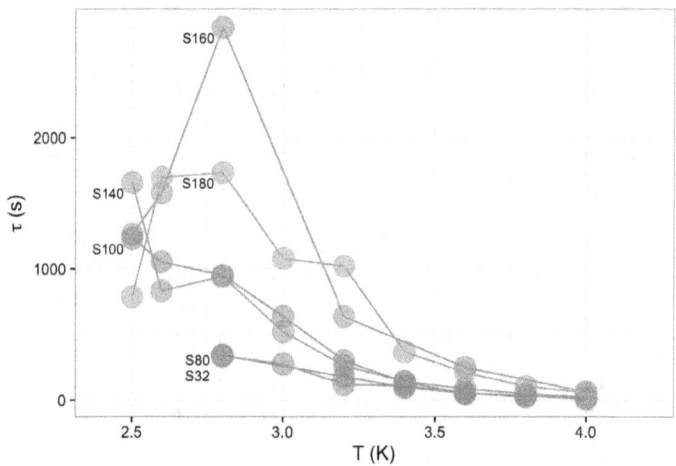

Figure 3.6: Relaxation times are fit from the stretched exponential model for all samples. Longer chains yield longer relaxation times. In this analysis, the values have not converged yet, and the time values are thought of as lower-bound limits.

Table 3.4: The relaxation time is fit for 8 samples at different measurement temperatures. Some fields are left empty, as the temperature during the measurement was not stable. All values are in units of s.

T (K)	S32	S80	S100	S140	S160	S180	S220	S230
2.5			1235	1658	1265	788		
2.6			1055	833	1580	1697		1776
2.8	345	338	951	942	2841	1735	1234	1780
3.0		277	639	522		1081	871	984
3.2	180	122	302	265	638	1023	487	467
3.4	99	110	140	145		370	248	214
3.6	55	56	50	87	252	214	152	111
3.8	27	36		52		107	93	76
4.0		12		23	64	58	58	39

whereas the times are much longer at lower temperatures. Secondly, the dynamic response appears to shape the behavior of the relaxation time somewhat. Samples S180 and S160 have a peak near 2.8 K, which could be expected from interdependence in the fit.

The frozen state, a metastability, in single chain magnets can be interpreted as an energy barrier ΔE that must be overcome to reach the full equilibrium state. The relaxation time in this kind of system changes rapidly with temperature, similar to spin glasses. Different expressions for the relaxation time dependence on temperature have been mentioned in the literature, including Vogel-Fulcher

$$\tau = \tau_0 \exp[A/(T - T_0)], \tag{3.2}$$

scaling expressions

$$\tau = \tau_0[(T - T_0)/T_0]^{-\gamma}, \tag{3.3}$$

and others.[25] Generally, for single chain magnets, an exponential, Arrhenius-like behavior is adopted[11, 75, 98]

$$\tau(T) = \tau_0(L) \exp\left[\frac{\Delta E}{k_B T}\right]. \tag{3.4}$$

In the finite-size regime, where the chain length L is less than the correlation length ξ, Glauber's dynamics with respect to infinitely long chains are modified and the single spin relaxation time τ_0 becomes chain length dependent. In the temperature range from 3 K to 4 K, the relaxation time τ appears to be activated, see Fig. 3.7. The slope is similar for all samples and roughly 3.2 meV. This means that $\Delta E/k_B \sim 37$ K. This should be compared with the nearest neighbor intra-chain interaction $J/k_B \sim 15$ K measured in FePc powder.[30] If the chain length L is shorter than the correlation length ξ, then the barrier height is $\Delta E = 2J$ rather than $\Delta E = 4J$ for Glauber's infinitely long chains. This would support our understanding that the chain length is an important driver of measured characteristics. The Arrhenius behavior for the relaxation time is fit to the following function (inverted from Eq. 3.4)

$$\frac{1}{\tau} = f_0(L) \exp\left[-\frac{\Delta E}{k_B T}\right]. \tag{3.5}$$

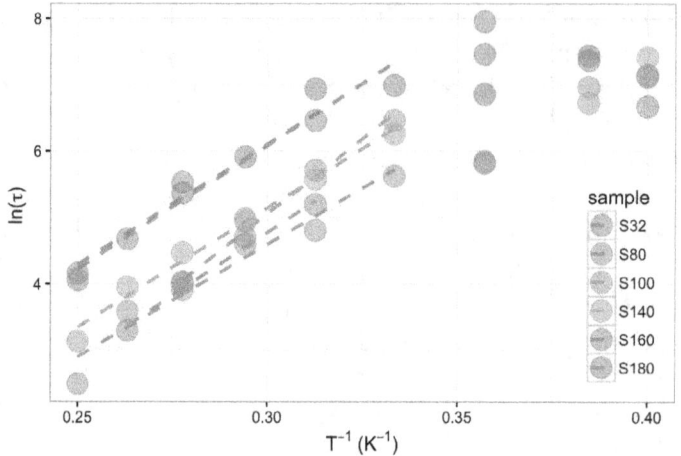

Figure 3.7: When the fitted relaxation time is plotted versus the inverse of the measurement temperature T, the temperature range from 3 K to 4 K shows an activation energy that is similar across all samples. The slope varies from 33 K^{-1} to 45 K^{-1}. Apart from sample S100, the samples have an approximate slope of 37 K^{-1}, which corresponds to ~3.2 meV. It is noted that the relaxation time τ is not fully relaxed in this analysis, see Tab. 3.8 for the relaxed values.

The results show that the characteristic frequency f_0 is around 1 kHz for samples with small crystals and drops to about 0.1 kHz for samples with larger crystals. In the work of Leal da Silva et al.[48] it is explained that in finite-sized chains the ends can flip at a higher rate as compared to the inner spins. This extra channel reduces the relaxation time, and they argue that the relaxation time τ_L of a finite-sized chain of length L increases with the chain length as $\tau_L \sim L\xi$.

The activation energy ΔE remains constant for all samples, see Tab. 3.5.

Another comparison is made with spin-glasses, which also show magnetic relaxation behavior, and follow the Vogel-Fulcher law.[82, 69] For spin-glasses, there is a transition temperature T_0. The viscosity η follows $\eta \sim \exp\left[-A/k_B(T - T_0)\right]$. Since viscosity $\eta \sim \tau$, measuring the relaxation time would provide insight whether it behaves similar to a spin glass, which would have $T_0 \neq 0$ K. For the presented data, the temperature range to probe is too narrow to have a meaningful discussion, however.

Table 3.5: The relaxation time is fit to $\tau^{-1} = f_0 \exp\left[-\Delta E / k_B T\right]$ from $3\,\mathrm{K}$ to $4\,\mathrm{K}$ as explained in the text to find the activation energy ΔE.

Sample	intercept	slope (K^{-1})	f_0 (Hz)	ΔE (meV)
S32	-6.6	38	735	3.3
S80	-5.6	34	270	2.9
S100	-8.5	45	4915	3.9
S140	-5.6	36	270	3.1
S160	-4.8	36	122	3.1
S180	-5.1	37	164	3.2
S220	-4.1	33	60	2.8
S230	-5.9	38	365	3.3

name	M_s ($\mu\,\mathrm{emu}$)	A (mm^2)	M_s/A ($\mathrm{n\,emu/cm}^2$)	M_s/V ($\mathrm{m\,emu/cm}^3$)
S32	150.7	33.98	44.36	2.77
S80	157.8	31.92	49.43	3.09
S100	148.3	30.45	48.70	3.04
S140	130.2	35.58	36.59	2.29
S160	161.7	35.28	45.83	2.86
S180	203.1	31.30	64.87	4.05
S220	185.0	33.62	55.04	3.44
S230	138.8	38.45	36.10	2.26

Table 3.6: Saturation moment M_s for each sample determined as an average measured at $2.5\,\mathrm{K}$ to $5\,\mathrm{K}$ assuming a common thickness of $160\,\mathrm{nm}$.

Table 3.7: Based on Gentry et al.[35] the deposition temperature T_{dep} controls the length of the major or long crystal axis d. The highest temperature denoted with * likely have chains that are broken and therefore the effective chain length is shorter than computed. The stretch exponent β is extracted from the master curve.

T_{dep} (°C)	d (nm)	β
32	39	0.26
80	48	0.17
100	56	0.16
140	73	0.17
160	90	0.20
180	107	0.25
220	168*	0.22
230	190*	0.19

3.6 Dependence on Average Grain Size

The sample series is distinguished by the deposition temperature. It has been shown that in MPc the grain size or crystal size is greatly affected by the deposition temperature.[35] In the following, we study the fit parameters in terms of the grain size, or crystal size. It should be mentioned that samples deposited above 200 °C have strongly increased roughness.[39] Although the exact lateral arrangement of standing MPc molecules on silicon substrates cannot be easily determined via x-ray diffraction in thin films, the atomic force microscopy images show distinct differences in samples deposited above 200 °C. For one, the mound formation gives way to terraced surfaces with many pin holes. Secondly, the grain size distribution is also affected. The tail in the distribution is longer.

Finite-size effects of the chain length have been at the core of several theoretical studies.[56, 48, 88] Experimental studies have not been able to alter the chain length systematically in single chain magnets, but the effective chain size has been reduced through impurities.[86, 87] The impurities have the effect of lowering the single spin relaxation time, however, the activation energy remains unchanged.

The deposition temperature is translated to an average chain length L using the data from Ref. [35]. The results of this conversion are tabulated in Tab. 3.7 for all samples. The grains are not symmetric and tend to grow in a particular direction, making the grains needle-like. In the analysis by Gentry et al. the grains are approximated with an ellipse, which has two axes. The important length is the major axis length, since the iron chains are directed along this direction. Between samples S32 and S180, the major axis length increases from 39 nm to 107 nm, its length more than doubles. The average length of the iron chains increases from about 100 ions to 300 ions of Fe(II). Since the spin of each molecule is $S = 1$, the net spin of a chain is equivalent to the number of iron ions per chain. As mentioned, samples S220 and S230 are deposited above the threshold temperature of 200 °C and therefore should be analyzed separately, since the grains are thought to have broken chains.

The longer iron chains lead to an exponential increase of the asymptotic magnetization fit value, see Fig. 3.8. The semi-log plot shows linear fits to the data points, which support this rapid increase with grain size. Lowering the temperature makes the effect more evident, at temperatures below 3.2 K, the slope D_0 of the fit varies between 30 nm to 34 nm, when expressed in the parametrized form of

$$M_\infty = A \exp\left[d/D_0\right], \tag{3.6}$$

where d is the length of the major axis. The amplitude A decreases from 13 µemu/cm^2 to 5 µemu/cm^2 as the temperature is changed from 2.5 K to 3.0 K.

The dynamic response ΔM equally increases strongly with the grain size, illustrated in Fig. 3.9. The maximum value depends on the measurement temperature.

The value of this analysis is limited as the fitting values are not converging. Therefore, the results should be interpreted qualitatively. In the following, the second analysis method is used. Rescaling of the temporal temperature-dependent remnance curves is used to extend the measurement time much beyond the instrument limited time of 5000 s. The rescaled curves are referred to as master curves. The stretch exponent is much smaller. In the first analysis $\beta = 0.43$, but using the master curves, we find $\beta \sim 0.2$, see Tab. 3.7. The value of the stretch exponent greatly affects the relaxation times, which in turn influence the values of the energy barrier and single spin relaxation time.

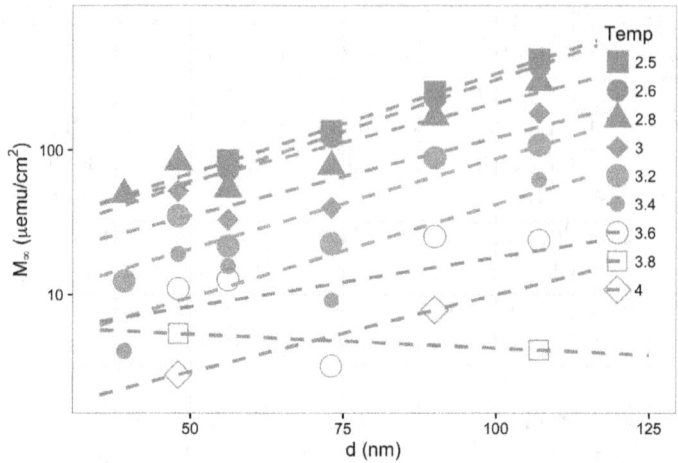

Figure 3.8: The asymptotic magnetization M_∞ increases exponentially with the grain size shown in the semi-log plot. The grain size is inferred from the deposition temperature. Samples deposited at temperatures above 200 °C are not plotted (see text) and their values for M_∞ are lower than sample S180. The dashed lines are fits to the data points.

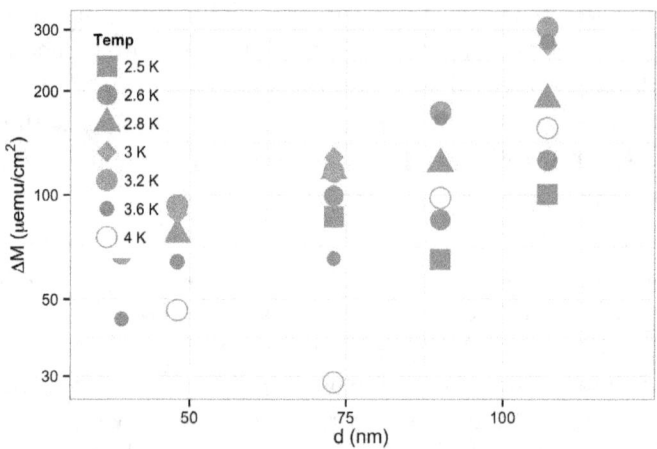

Figure 3.9: The dynamic response ΔM is graphed for 5 different samples of FePc/Si. The deposition temperature is converted to the major axis length of the grain size d. The increase of grain size yields a stronger dynamic response.

3.7 Master Curve Method

Due to the convergence issues of the previous method, we will use an analysis whereby temperature-independent master curves are generated from the temporal remanent curves. These master curves span many decades and can be fit more reliably. In powder samples, it has been shown that this method yields results which are consistent with ac susceptibility measurements.[31]

In the single chain magnet, $(NEt_4)[Mn_2(5\text{-MeOsalen})_2Fe(CN)_6]$, with $5\text{-MeOsalen}^{2-} = N,N'\text{-ethylenebis}(5\text{-methoxysalicylideneiminate})$, made with repeated Mn(III)-Fe(III)-Mn(III) chain elements, the remnant magnetization is normalized by the relaxation time τ to generate a master curve, for instance. This system was studied in exhaustive detail. The traditional frequency-dependent susceptibility measurements are extended at low-frequencies with temporal remanent magnetization measurements based on dc measurements.

For thin film samples deposited onto a carrier layer, the susceptibility measurements become complex, as the signal is convoluted. As pointed out in the previous section, the analysis of the temporal remanent magnetization requires very long measurements ($t \gg 5000\,\text{s}$). In order to circumvent the instrument limitations, a temperature-independent master curve is generated from both low-temperature and high-temperature measurements. The underlying idea is that the relaxation is sped up at higher temperatures. By overlapping data measured at successive temperatures, a scaling factor is extracted. In this approach, the data is first normalized by the saturation moment M_s, see Fig. 3.10 for sample S180.

The master curve is generated from data measured below 4.1 K. Starting with the lowest temperature, the data from the higher temperature is matched at two points and the scaling ratio is determined. An analysis of this procedure shows that choosing the particular points does not modify the χ^2 value between the two curves. Ideally, a full overlap is expected. The largest discrepancy in the matched up curves occurs in the first few data points. Repeating the procedure for all temperatures, a single master curve is obtained, see Fig. 3.11. The curve now includes 8 decades of time and would be equivalent to measuring the sample over a period of about 3 years at the lowest measured temperature; i.e. at 2.5 K. This procedure is repeated for all samples.

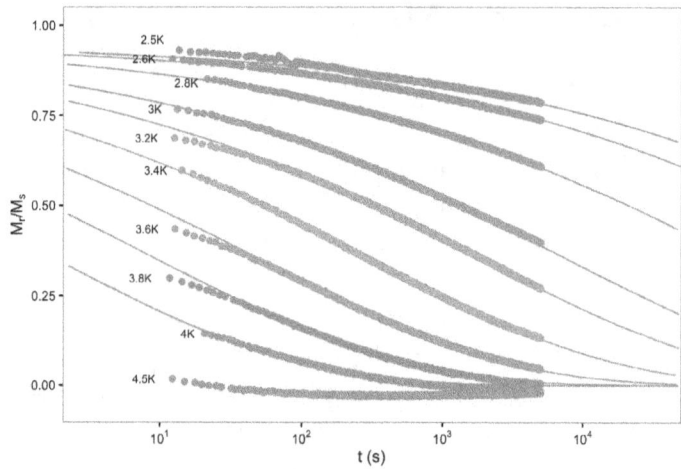

Figure 3.10: Temporal remnance curves measured for sample S180 in the range of 2.5 K to 5 K. At low temperatures, the relaxation is very slow. The process is accelerated successively at higher temperatures. The lines are the fits obtained from the master curve.

From the master curve the stretching exponent is extracted. In this case, β varies from 0.16 to 0.26. The low value seems to indicate that the sample is quite polydisperse. The best fit values of β for all samples are summarized in Tab. 3.7.

From the master curve, any particular temperature measurement is extrapolated in time using this stretched exponent. These fit lines capture all temperature measurements quite well, see Fig. 3.10. This now yields the relaxation time $\tau(T)$ for each measured temperature.

From the Glauber-Ising model, we would expect an Arrhenius law behavior of the relaxation time, see Eq. 3.4. Indeed, when plotting the temperature-dependent relaxation times for all samples, an activation energy barrier is extracted, see Fig. 3.12. Deviations from this behavior are seen for the shortest chains and as the temperature approaches 4 K.

As observed for single chain magnets, the extracted relaxation time with temperature has a characteristic energy barrier height Δ_A. In the form graphed in Fig. 3.12, the slope represents the energy barrier height. The data suggest that the slope is independent of the chain length L. In the Glauber-Ising model, the energy barrier is interpreted as the energy needed to flip a spin. Applying fits to the low-temperature data ($T < 3.4$ K), the barrier height is determined as

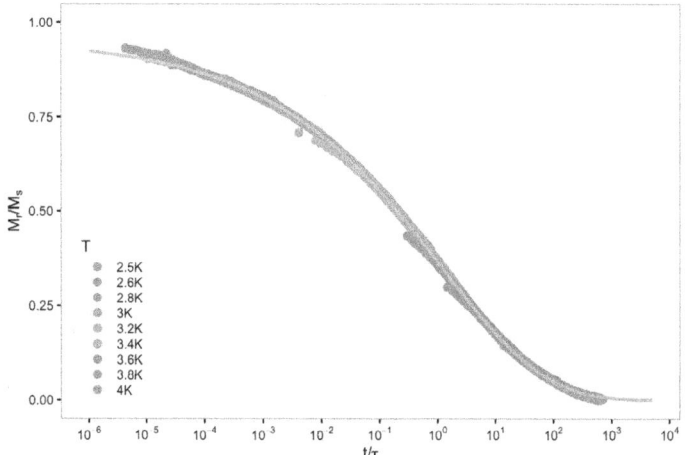

Figure 3.11: The master curve for sample S180. Each temperature measurement is scaled and fit together to form a temperature-independent master curve. Since the master curve spans almost 8 decades in time, it can be fit with a stretched exponential function that has a small β parameter. For the shown fit, $A = 0.95$, $\tau = 1.21$ s, and $\beta = 0.25$.

Table 3.8: Fitting parameters for sample S180 are extracted from the master curve with fixed $\beta = 0.25$. The amplitude A of the remanent magnetization is constant in this temperature range. The relaxation time τ increases rapidly as the temperature is lowered.

T	A	τ
(K)	(μ emu)	(s)
2.51	161	3.85×10^6
2.60	161	1.34×10^6
2.80	161	1.28×10^5
3.00	161	8.02×10^3
3.20	161	1.97×10^3
3.40	161	307
3.60	161	52.2
3.80	161	9.70
4.00	161	1.81

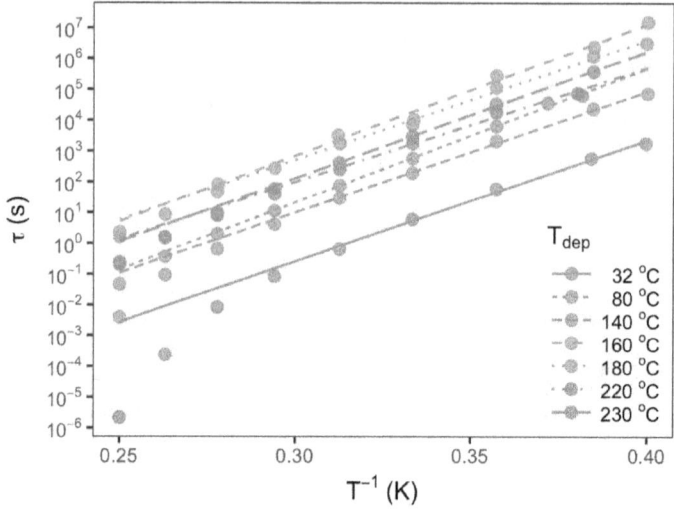

Figure 3.12: The relaxation times are determined from fits to $M_r(t)$ using the stretched exponential form. The best fit lines are linear fits to extract the energy barrier Δ_A and the intercept τ_0.

$\Delta_A/k_B = 95(\pm 4)$ K, see Fig. 3.13. For single chain magnets, this barrier height is proportional to the anisotropy energy D.[19] It remains constant for all FePc samples. For samples deposited under 200 °C, the AFM crystal size and coercivity increase monotonically. For the highest deposition temperatures, the coercivity decreases, even though AFM images show the longest crystals. It suggests that the crystals are limited by defects. Further evidence comes from Tab. 3.3 showing the parametrized amplitude of the dynamic response, which decreases after reaching a peak for the S180 sample. The chain length in two samples deposited at high temperatures are estimated from coercivity data instead of AFM images. This results in nominal chain lengths between 80 nm and 100 nm. The data points are added to the graph in Fig. 3.13 with open symbols to distinguish the low-temperature samples, for which the grain size is determined via atomic force microscopy (AFM).

As a reference, we estimate the correlation length and the associated energy barrier. Using data from magnetic circular dichroism measurements on FePc/Au/Si samples,[37] we can estimate Δ_ξ using the relation that $\chi T \sim \exp(\Delta_\xi/k_B T)$ and fit the high-temperature data.

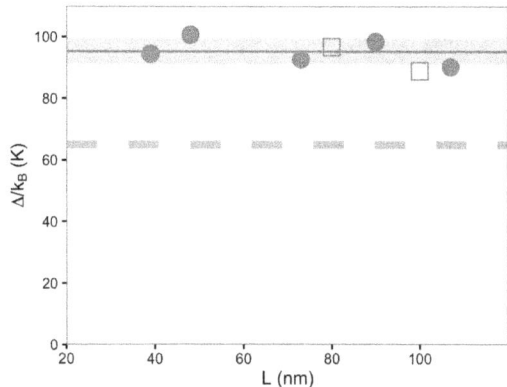

Figure 3.13: Energy gap Δ_A/k_B versus chain length of sample. The gray line is added for reference for $\Delta_\xi/k_B = 64$ K determined from MCD data in FePc/Au thin films. The chain length of 5 samples (circle) is determined from the crystal size using AFM, the other two samples (square) are inferred lengths from coercivity measurements.

Using MCD data,[37] we find $\Delta_\xi = 64$ K, which allows us to estimate the correlation length ξ given the form $2\xi/a = \exp(\Delta_\xi/k_BT)$.[20] The distance a corresponds to the Fe-Fe distance. We conclude that $\xi \gg L$ for all measurements below 5 K. As an example, at 4 K, we arrive at $\xi \sim 1.5 \times 10^6$ nm. Given the different geometry of FePc/Si thin films, this solely serves as an estimate.

In the Glauber-Ising model, the single spin relaxation time τ_0 is an adjustable parameter with a dependence on the chain length, if $\xi \gg L$. Experimentally, the spin relaxation time corresponds to the y-intercept in Fig. 3.12 and varies with chain length. Small variations in the slope, however, amplify the intercept position. Therefore, the measured single-spin relaxation times τ_0 should be interpreted with caution. We find a strong dependence of τ_0 with the chain length L as summarized in the graph of Fig. 3.14. The results are consistent with the impurity-doped crystals that show diminished values for τ_0 as impurities are introduced and the findings agree with theoretical results of finite-sized magnetic chains that argue the single spin relaxation time is correlated with chain length, if the condition $\xi \gg L$ is met.[48, 56, 86] Still, finite-size Ising chains are predicted to have a linear dependence on L, slower than our observations suggesting that the FePc thin film system may not fulfill all assumptions in the theoretical framework.

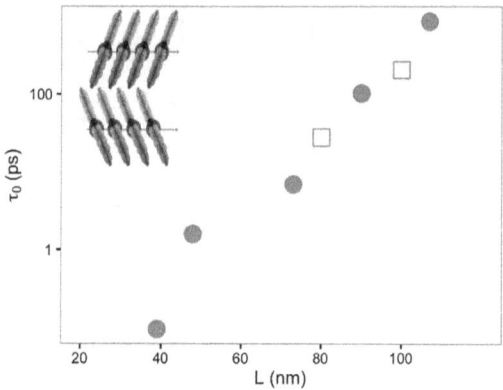

Figure 3.14: Single spin relaxation times for 7 samples with different chain lengths. For two samples (squares) deposited above 200 °C the length L is estimated indirectly. Inset shows a top view of the herringbone arrangement of FePc molecules, the arrow indicates the direction of the magnetic field and the direction of the b-axis.

Indeed, the sub-lattice magnetization of a herringbone structure as shown in the inset of Fig. 3.14 with an xy anisotropy in the plane of the molecule may result in more complex magnetic dynamics. This zig-zag structure essentially cancels one component, so that the net magnetization points along the b-axis, which is averaged over the plane.

In summary, low-dimensional magnetic spin systems play a key role as paradigms for understanding dynamic magnetic behavior. The Glauber-Ising model has been extended to capture dynamic effects of finite-sized systems theoretically,[48, 56, 86] but experimental realizations in powder and crystal samples suffer from random defects that prevent studying the explicit chain length dependence. In this chapter, we built nano-structures via the control of thin film deposition parameters to vary the chain length systematically. The relaxation time of FePc thin films is extracted using dc measurements and shows strong dependence on the chain length L in the regime of $\xi \gg L$. We find using the second analysis method that the energy barrier Δ_A is independent of deposition temperature and chain length. Using a simple molecule, iron phthalocyanine, we demonstrate the effect of tuning the chain length to achieve control of the magnetic relaxation time. Using templated substrates and careful growth conditions, the chain lengths in FePc thin films can be designed to achieve locally variable magnetic relaxation times. The system also provides a useful experimental realization of tunable finite-sized chain systems and experimental insight into magnetic dynamics of finite-sized ion chains, which are broadly accessible with theoretical tools.

3.8 Iron Phthalocyanine on Gold Surface

The structural configuration of FePc on smooth gold surfaces is different from that on silicon.[71] In order to probe the relaxation behavior in different structures, two samples were co-deposited onto silicon and gold/silicon surfaces respectively. The samples are listed as A180 and S180 on page 26, each has FePc deposited at 180 °C with the same thickness of 154 nm. Since the samples were deposited at the same time on a rotating deposition plate, and the magnetic properties were characterized close in time, a direct comparison is insightful. The expectation is that the chains in the A180 sample are growing perpendicular to the substrate surface, which has been corroborated elsewhere via x-ray diffraction.[37] However, it is not clear that it holds true for large thicknesses. It is possible that the initial layers grow in the Lying configuration, while successive layers revert to the Standing configuration at some thickness.

The applied field will be perpendicular to the chain axis for the samples with FePc deposited onto Au. The chains are not 154 nm long, but rather broken at lengths below this maximum length limit.[93]

The full set of temporal remnant magnetization curves for A180 is shown in Fig. 3.15. Comparing the temporal evolution with Fig. 3.3, qualitatively similar trends are observed. At 4.5 K there is almost no remnant magnetization, however, at 4.0 K the dynamic response is more than 200 µemu/cm^2, see Tab. 3.10. At the lowest temperature of 2.6 K the relaxation is very slow and still incomplete after 5000 s. A side by side comparison shows subtle differences. In Fig. 3.16 the temporal remnant magnetization is graphed along with fits to the measured data. At the measured temperature 3.4 K the dynamic response of A180 is slightly smaller compared with S180; in numbers, $\Delta M = 306$ µemu/cm^2 for A180 compared with $\Delta M = 325$ µemu/cm^2 for S180. Even more pronounced is the difference in relaxation times. Sample S180 has relaxation times that peak at lower temperatures and are longer. A full comparison of the relaxation times is provided in Tab. 3.10.

From the previous analysis, slower relaxation time and higher tem-

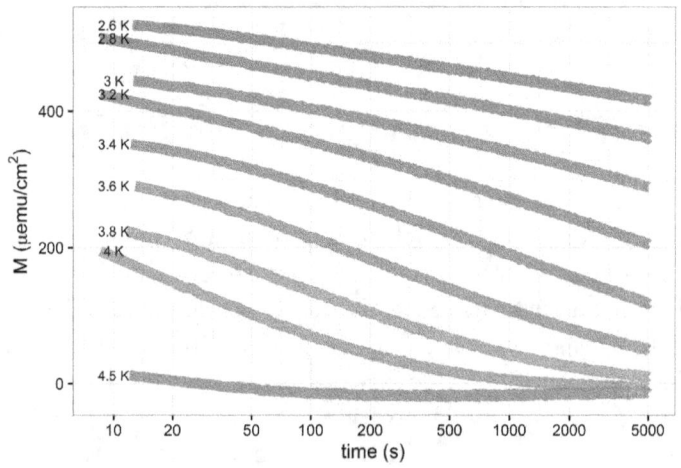

Figure 3.15: Remnant magnetization measured at nine different temperatures for sample A180 shows relaxation for $T < 4.5\,\mathrm{K}$. A180 is FePc deposited on Au/Si. The relaxation can be compared to the co-deposited sample S180 that is deposited on Si instead.

T (K)	M_∞ A180 (μemu/cm^2)	M_∞ S180 (μemu/cm^2)
2.5		419
2.6	405	377
2.8	348	294
3	261	180
3.2	171	108
3.4	88	62
3.6	39	24
3.8	10	4
4	-4	-7
4.5	-16	

Table 3.9: Asymptotic magnetization fitting values are compared for A180 and S180. Both samples were deposited at the same time onto different surfaces.

T (K)	ΔM A180 (μemu/cm^2)	ΔM S180 (μemu/cm^2)	τ A180 (s)	τ S180 (s)
2.5		101		788
2.6	138	126	683	1697
2.8	163	189	704	1735
3.0	204	271	1120	1081
3.2	266	304	1040	1023
3.4	306	325	795	370
3.6	312	278	376	214
3.8	290	225	160	107
4.0	288	156	51	58
4.5	182		3	

Table 3.10: Dynamic response ΔM and relaxation time τ fitting values in comparison for samples A180 and S180.

perature peak for ΔM suggest that the A180 sample has shorter chains.

The most striking difference between the two samples deposited onto different substrates is possibly the peak position of the dynamic response. For both samples, the peak magnitude of the dynamic response are similar at 320 μemu/cm^2 and 315 μemu/cm^2. However, the peaks are shifted in position as shown in Fig. 3.17. For the sample deposited onto a gold surface with the iron chains growing perpendicularly to the substrate the peak appears at a higher temperature, near 3.62 K. This position is higher than any other measured sample, cf. Tab. 3.3 on page 47. The peak for A180 is also broader.

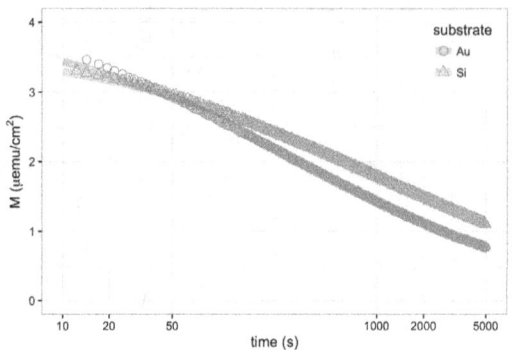

Figure 3.16: Samples A180 and S180 were co-deposited onto Au/Si and Si substrates, respectively. A comparison of the measurements at 3.4 K show that the remnant magnetization relaxes differently. The asymptotic magnetization M_∞ is $0.64\,\mu\text{emu/cm}^2$ and $0.82\,\mu\text{emu/cm}^2$ for A180 and S180 respectively. The change in magnetization is stronger in A180 at $3.43\,\mu\text{emu/cm}^2$ compared to $2.9\,\mu\text{emu/cm}^2$ for S180, and the relaxation for S180 is about double with 800 s versus 390 s.

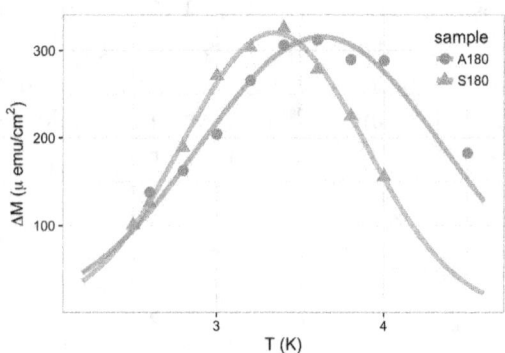

Figure 3.17: Dynamic response is graphed versus measurements temperature for two co-deposited samples with different substrates. The lines are Gaussian fits tracking the measurements. The amplitudes for both fits are similar at $320\,\mu\text{emu/cm}^2$, but the peak positions are shifted. For the sample deposited onto gold, the peak is near 3.62 K much higher than the peak for S180 at 3.34 K. The widths are 0.54 K and 0.72 K for S180 and A180, respectively.

4 Finite-Sized Iron Ion Chains

4.1 Summary

Ferromagnets are widely used in modern applications such as data storage, audio devices, switches, and transformers. A search for novel materials is on-going to either replace costly strong ferromagnets or extend optical properties. This would open the way for new applications. Novel systems based on organometallics include single molecule magnets and single chain magnets and exhibit magnetic hysteresis due to slow relaxation explained by Glauber dynamics. Metallo-organic materials are low density, low weight, and flexible. They can be solution processed and have unique optical properties suited for new tasks and technologies.

The magnetic properties of iron phthalocyanine have been probed in powder about 16 years ago. Evangelisti et al. measured the remnant magnetization in FePc powder and found that from 1.8 K to 5 K the value dropped rapidly.[30] Above 5 K, however, the decay of the remanence was much slower and a small amount of remnant magnetization persists up to 15 K. The observation of non-zero values of remanence after the rapid decrease in remanence is quite unusual. Concretely, the remanence at 2 K is almost 800 emu/mol and drops to just below 20 emu/mol at 5 K. Magnetic hystersis can only be observed in loops below 5 K. This behavior was interpreted in terms of a strong ferromagnetic short-range order along the chains at temperatures below 15 K. From the susceptibility data a inter-chain interaction strength J/k_B of 15 K was extracted. Susceptibility data also shows strong frequency dependence. The frequency dependence of the freezing temperature T_f was determined for different frequencies. The shift is measured in terms of the ratio $(\Delta T_f/T_f)/\Delta \ln(\omega)$ with a value of 0.034. The authors found that it is 6 times larger than a metallic spin glass, but also smaller than a superparamagnet, see page 16.

More recently, interest has turned to FePc thin films.[38] Due to the van der Waals forces between molecules, the arrangement in thin films can be varied with the deposition temperature and substrate choice. Here, a series of eight 154 nm thick FePc samples on silicon substrates are studied. Additionally, one sample deposited on gold-coated silicon is compared with its counterpart deposited onto Si only. The deposition temperature is varied from 32 °C to 230 °C. The purpose of the different deposition temperatures is to increase the average chain length L of the sample. FePc thin films are comprised of crystals with needle-like shape. These crystals self-assemble so that Fe(II) ions form chains. Super-exchange along Fe-N-Fe bonds provides the mechanism for spin ordering along the chain direction. The iron chains form along the b-axis, which runs either parallel to the substrate plane for FePc/Si, or perpendicular to the substrate for FePc/Au/Si samples.

All thin films were deposited in high-vacuum using a thermal evaporator ($< 10^{-6}$ mbar). The thickness was determined from x-ray diffraction in one of the samples. It was determined that the thickness is 154 nm using a fringe oscillation next to the main peak at $2\theta = 6.93°$. The main peak position corresponds to a d-spacing of 12.7 Å consistent with literature values for FePc standing on the substrate and having the b-axis parallel to the substrate.

The magnetic remnant state is investigated as a function of the crystal size. The crystal size is related to the average iron chain length. At first, each sample is saturated at 3 T before measurements at different temperatures between 2.5 K to 5 K. The magnetic field is reduced to zero in order to record the temporal behavior of the remnant magnetization. The speed to remove the magnetic field is 106 Oe/s limited by the instrument. Each sample is measured during a period of 5000 s in zero magnetic field. In agreement with findings for FePc powder, the remanence almost completely vanishes at temperatures above 4.5 K. For lower temperatures, however, the temporal remnant magnetization can be fit to a stretched exponential function (Eq. 1.25).

Two methods were applied. In the first method, three independent parameters are fit, while the stretch exponent β is fixed at the fraction 3/7. The independent parameters are the asymptotic magnetization M_∞, the dynamic response ΔM, and the relaxation time τ. In the second method, remanence curves from several temperatures are collapsed into one master curve, which could be fit to determine the stretch exponent. The exponent β turned out to be between 0.16 and 0.26, very low and indicative for polydisperse samples.

The measurement period of 5000 s was instrument limited. The temperature could not be stabilized routinely for longer periods. An investigation of the robustness of the fitting parameters showed on page 33 that the fit parameters do not yet converge fully. For time spans above 2000 s the fit parameters, however, have clear trends. In particular, the relaxation time increases and the asymptotic magnetization decreases with longer measurement periods.

Each fit parameter has been analyzed. The asymptotic magnetization M_∞ is non-zero for all samples at sufficiently low temperatures. It vanishes before reaching the 4.5 K temperature where hysteresis loops are measured. Since the fit values are expected to decrease with longer measurement times, the asymptotic magnetization values should be considered an upper limit. Their values measure the coupling between iron chains. The asymptotic magnetization increases with grain size exponentially at temperatures below 3.2 K suggesting that the relaxation is slower and energy barriers are higher in samples with longer iron chains.

The dynamic response ΔM relates to the amount of relaxation that occurs. This fitting value has a temperature dependence that follows a Gaussian form, reminiscent of the out-of-phase susceptibility dependence with temperature. It goes to zero at 4.5 K and low temperatures, but it has a peak at intermediate temperatures. The peak position depends on the sample preparation conditions. It increases with iron chain length. For the sample deposited at 32 °C the peak is near 2.7 K and shifts towards 3.3 K for the sample with the longest iron chains, sample S180.

The relaxation time τ is largest for lower temperatures and it increases with iron chain lengths, for samples grown at higher deposition temperatures. The relaxation times follow an Arrhenius law in the small temperature range 3 K to 4 K with an energy barrier height of 37 K. This barrier height is attributed to the frozen state of the magnetization and provides a measurement for the product of the coupling strength in a single chain magnet. It is consistent with the fact that the chain length L is smaller than the correlation length ξ and the chains must be considered finite in length. The modification of the kinetic Glauber model implies a chain length dependent spin relaxation time $\tau_0(L)$. The open ends tend to increase the chance for flipping. The experiment shows an increase of relaxation time with chain length as graphed on page 60.

Since the 5000 s measurement time is insufficient for full convergence of the data, the temporal remanence measurements at higher temperatures are rescaled to append the low temperature measurement. The resulting master curve is fit with a stretched exponential function to determine the stretch exponent. Once known, each temperature measurement can be fit for the relaxation time using a single fit parameter. It turns out that the relaxation time follows an Arrhenius law (page 58) as expected for a single chain magnet. The slopes for all samples are similar and indicate a single activation energy of 95 K. This suggests that the interaction strength D is constant and material specific. In contrast, the single spin relaxation time increases rapidly from less than 1 ps to several hundreds of pico seconds for the longest chains. Since the longer chains have fewer end chain spins, the relaxation is expected to be slower.

Two co-deposited samples with different substrates are compared side-by-side. Since the b-axis is perpendicular to the substrate on the gold samples, the applied field is now perpendicular to the axis. Qualitatively, similar relaxation behavior is observed. All temporal curves can be fit to a stretched exponential function. A comparison at the measurement temperature of 3.4 K shows that the relaxation time is almost twice as long for the sample deposited onto silicon.

We conclude the remnant magnetization of iron phthalocyanine slowly relaxes below 4.5 K. The relaxation can be fit to a stretched exponential function with an exponent below the fraction of $3/7$. The relaxation is very slow in this metallo-organic thin film. Indeed, it appears to have a broad distribution, such that the experimental measurement time of 5 ks is not sufficiently long for fitting values to converge. Still, using a rescaling approach, master curves can be generated for all samples. It is suggested that the interaction between Fe(II) ions is mediated via superexchange, resulting in weakly ferromagnetically chains. The chains are separated by about 1.3 nm with a lack of π-conjugated links, suggesting that chains are magnetically insulated. The intra-chain interactions are much stronger as compared to the magnetic inter-chain interactions. The intra-chain coupling can be estimated from the activation energy determined from the temperature-dependent relaxation time. A constant value of 95 K is extracted from 8 samples with different chain lengths. However, the single spin relaxation time τ_0 rapidly increases with chain length. This increase is much faster than linear as had been suggested on the basis of a theoretical model. Clearly, the sample measurements include ensembles and wide distributions of chain length. Some impurities are also present. Regardless, the deposition temperature provides a useful knob in controlling the coercivity and single relaxation time of FePc thin films. It presents a way forward to studying finite-sized magnetic ion chains, or finite-sized single chain magnets.

5 Ising Model

5.1 Correlation Length in 1D Ising Model

The 1D Ising model has simplicity, yet provides extremely rich behavior. It has educational value and applications beyond magnetism. In one dimension, analytical solutions can easily be derived for a magnetic coupled linear chain as shall be shown here. The Ising model was first proposed by Ernst Ising's Ph.D. advisor Wilhelm Lenz and published in the Zeitschrift für Physik in 1925.[44] It shows that this one-dimensional model shows no ferromagnetic-paramagnetic transition. Rudolf Peierls later referred to Ising's model in a publication in the Mathematical Proceedings of the Cambridge Philosophical Society, where he considers the 2D case.[70] In 1944, Lars Onsager solved the model in two dimensions analytically.

In the following, we show that there is no phase transition in the 1D Ising model by calculating the correlation length ξ. Here, we consider a lattice with N sites, each having a spin S_i, where i denotes the location in the lattice. In the Ising model, the spin can only assume two values ± 1. The interaction energy has two terms, in the first term the spin interaction is considered and the second term includes the Zeeman term.

$$E = -\sum_{i,j} J_{ij} S_i S_j - \sum_{i=1}^{N} B_i S_i \qquad (5.1)$$

Immediately, we will make some simplifications and only assume nearest neighbor interactions. Therefore, the J_{ij} reduces to a single value J, which is positive for ferromagnetic interaction and negative for antiferromagnetic interactions. We will assume a constant magnetic field B at each site location. We will then write the partition function

$$Z = \prod_{i=1}^{N} \sum_{S_i=-1}^{1} e^{-\beta E}, \qquad (5.2)$$

where $\beta = (k_B T)^{-1}$. Kramers and Wannier showed that the partition can be solved using a simple matrix P. They defined the following matrix

$$P = \begin{pmatrix} e^{\beta(J+B)} & e^{-\beta J} \\ e^{-\beta J} & e^{\beta(J-B)} \end{pmatrix} \tag{5.3}$$

then it follows that

$$\langle S|P|S' \rangle = \exp\left(\beta \left[JSS' + \frac{1}{2}B(S + S') \right] \right) \tag{5.4}$$

therefore the partition function Z is now expressed as follows

$$\sum_{S_1=-1}^{1} \cdots \sum_{S_N=-1}^{1} \langle S_1|P|S_2 \rangle \langle S_2|P|S_3 \rangle \ldots \langle S_N|P|S_1 \rangle = \sum_{S_1=-1}^{+1} \langle S_1|P^N|S_1 \rangle \tag{5.5}$$

with the result being the trace of the matrix, so that the solutions are two eigenvalues, namely

$$Z = Tr\left(P^N \right) = \lambda_+^N + \lambda_-^N \tag{5.6}$$

The eigenvalues are computed by solving $Px = \lambda x$, where x is the eigenvector. Computing the determinant of $(P - \lambda I)$, where I is the identity matrix, provides the following two solutions

$$\lambda_{\pm} = e^{\beta J} \left[\cosh(\beta B) \pm \sqrt{\cosh^2(\beta B) - 2e^{-2\beta J}\sinh(2\beta J)} \right] \tag{5.7}$$

which can be simplified in the case for $B = 0$. In the thermodynamic limit ($N \to \infty$), only λ_+ is relevant. Having solved the partition function, allows us to compute the free energy and the magnetization. The free energy is related to the quantity $\ln Z$ and turns out to be $F = -Nk_B T \ln \lambda_+$, since $\lambda_- \ll \lambda_+$. The sample magnetization M is computed to be

$$M = \frac{1}{\beta}\frac{\partial \ln Z}{\partial B} = -\frac{\partial F}{\partial B} = \frac{\sinh(\beta B)}{\sqrt{\cosh^2(\beta B) - 2e^{-2\beta J}\sinh(2\beta J)}} \tag{5.8}$$

If we set the magnetic field to 0 in this equation, then $m = 0$ for all temperatures. This means that there is no spontaneous magnetization in the 1D infinitely long Ising model. The susceptibility χ can also be calculated, since $\chi = \partial M / \partial B$.

Another quantity of interest is the spin-spin correlation length. This quantity measures the interaction of spin S_i with another spin S_j. This statistical measurement lends the connection with the correlation length ξ,

$$\langle S_i S_j \rangle - \langle S_i \rangle \langle S_j \rangle \sim \exp\left(-\frac{|i-j|}{\xi}\right) \tag{5.9}$$

It follows that this quantity vanishes, if the spins are completely independent. Typically, this measure decays exponentially with distance ($|i-j|$). Therefore, we can introduce a quantity called the correlation length ξ. The latter quantity can be measured and we can talk about weakly and strongly correlated systems.

For the 1D Ising model, the correlation length is computed from the spin correlations,

$$\langle S_k S_l \rangle_N = \frac{1}{Z} \prod_{i=1}^{N} \sum_{S_i=-1}^{1} S_k S_l e^{-\beta E} = \frac{\lambda_+^{N-(l-k)} \lambda_-^{l-k} + \lambda_+^{l-k} \lambda_-^{N-(l-k)}}{\lambda_+^N + \lambda_-^N} \tag{5.10}$$

In the thermodynamic limit, this simplifies to

$$\lim_{N \to \infty} \langle S_k S_l \rangle_N = \left(\frac{\lambda_-}{\lambda_+}\right)^{l-k} \tag{5.11}$$

let us set $r = |l-k|$, and have $B = 0$, then

$$\langle S_k S_l \rangle_{B=0} = (\tanh(\beta))^r \tag{5.12}$$

Thus, solving for ξ in Equation 5.9

$$\xi_{1D} = \frac{-1}{\ln(\tanh(\beta))} \tag{5.13}$$

Consequently, as temperature is reduced towards absolute zero ($\beta \to \infty$), the correlation length ξ diverges.

5.2 Critical Exponents for 2D Ising Model

The results for the 1D case are quite different from a 2D plane. According to Onsager, in two dimensions, near the critical point T_c, the spin-spin correlations decay asymptotically as a power law. The critical exponent (2D Ising) is $\nu = 1$.

$$\xi_{2D} \sim \left| \frac{T - T_c}{T_c} \right|^{-\nu} \tag{5.14}$$

The other critical exponents in 2D Ising model are the dependence of the heat capacity, $c_V \sim |T - T_c|^{-\alpha}$ with $\alpha = 0$. The magnetization near the critical point can be scaled with $M \sim |T - T_c|^{\beta}$, where $\beta = 1/8$. The magnetic susceptibility follows $\chi \sim |T - T_c|^{-\gamma}$ with $\gamma = 7/4$. As mentioned, these critical exponents are known exactly.

List of Tables

List of Figures

Bibliography

[1] Ao, R., Kümmerl, L., and Haarer, D. Present limits of data storage using dye molecules in solid matrices. *Advanced Materials 7*, 5 (May 1995), 495–499.

[2] Bartolomé, F., Bunău, O., García, L. M., Natoli, C. R., Piantek, M., Pascual, J. I., Schuller, I. K., Gredig, T., Wilhelm, F., Rogalev, A., and Bartolomé, J. Molecular tilting and columnar stacking of Fe phthalocyanine thin films on Au(111). *Journal of Applied Physics 117*, 17 (May 2015), 17A735.

[3] Bartolomé, J., Bartolomé, F., Figueroa, A. I., Bunău, O., Schuller, I. K., Gredig, T., Wilhelm, F., Rogalev, A., Krüger, P., and Natoli, C. R. Quadrupolar XMCD at the Fe K-edge in Fe phthalocyanine film on Au: Insight into the magnetic ground state. *Physical Review B 91*, 22 (June 2015), 220401.

[4] Bartolomé, J., Bartolomé, F., García, L. M., Filoti, G., Gredig, T., Colesniuc, C. N., Schuller, I. K., and Cezar, J. C. Highly unquenched orbital moment in textured Fe-phthalocyanine thin films. *Physical Review B 81*, 19 (May 2010), 195405.

[5] Berkov, D. V., and Kötitz, R. Irreversible relaxation behaviour of a general class of magnetic systems. *Journal of Physics: Condensed Matter 8*, 9 (1996), 1257.

[6] Bhattacharjee, S. M., and Khare, A. Fifty Years of the Exact Solution of the Two-Dimensional Ising Model by Onsager. *arXiv:cond-mat/9511003* (Nov. 1995).

[7] Blundell, S. J., and Pratt, F. L. Organic and molecular magnets. *Journal of Physics: Condensed Matter 16*, 24 (2004), R771.

[8] BONNER, J. C., AND FISHER, M. E. Linear Magnetic Chains with Anisotropic Coupling. *Physical Review 135*, 3A (Aug. 1964), A640–A658.

[9] BRINKMANN, H., KELTING, C., MAKAROV, S., TSARYOVA, O., SCHNURPFEIL, G., WÖHRLE, D., AND SCHLETTWEIN, D. Fluorinated phthalocyanines as molecular semiconductor thin films. *physica status solidi (a) 205*, 3 (Mar. 2008), 409–420.

[10] BROWN, W. K., AND WOHLETZ, K. H. Derivation of the Weibull distribution based on physical principles and its connection to the Rosin–Rammler and lognormal distributions. *Journal of Applied Physics 78*, 4 (Aug. 1995), 2758–2763.

[11] CANESCHI, A., GATTESCHI, D., LALIOTI, N., SANGREGORIO, C., SESSOLI, R., VENTURI, G., VINDIGNI, A., RETTORI, A., PINI, M. G., AND NOVAK, M. A. Glauber slow dynamics of the magnetization in a molecular Ising chain. *Europhysics Letters (EPL) 58*, 5 (June 2002), 771–777.

[12] CANESCHI, A., GATTESCHI, D., SESSOLI, R., BARRA, A. L., BRUNEL, L. C., AND GUILLOT, M. Alternating current susceptibility, high field magnetization, and millimeter band EPR evidence for a ground S= 10 state in [$Mn_{12}O_{12}$ ($CH_3COO)_{16}(H_2O)_4$]. $2CH_3$ COOH. $4H_2O$. *Journal of the American Chemical Society 113*, 15 (1991), 5873–5874.

[13] CANNELLA, V., AND MYDOSH, J. A. Magnetic Ordering in Gold-Iron Alloys. *Physical Review B 6*, 11 (Dec. 1972), 4220–4237.

[14] CARDONA, M., CHAMBERLIN, R., AND MARX, W. The history of the stretched exponential function. *Annalen der Physik 16*, 12 (Dec. 2007), 842–845.

[15] CHAMBERLIN, R. V. Time-decay of the thermoremanent magnetization in spin-glasses as a function of the time spent in the field-cooled state. *Physical Review B 30*, 9 (Nov. 1984), 5393–5395.

[16] CLÉRAC, R., MIYASAKA, H., YAMASHITA, M., AND COULON, C. Evidence for Single-Chain Magnet Behavior in a Mn^{III}-Ni^{II} Chain

Designed with High Spin Magnetic Units: A Route to High Temperature Metastable Magnets. *Journal of the American Chemical Society 124*, 43 (Oct. 2002), 12837–12844.

[17] COLESNIUC, C. *Metallophthalocyanine Thin Films : Structure and Physical Properties*. PhD thesis, University of California San Diego, 2011.

[18] COOK, M. J. Phthalocyanine thin films. *Pure and Applied Chemistry 71*, 11 (1999), 2145–2151.

[19] COULON, C., CLÉRAC, R., LECREN, L., WERNSDORFER, W., AND MIYASAKA, H. Glauber dynamics in a single-chain magnet: From theory to real systems. *Physical Review B 69*, 13 (Apr. 2004), 132408.

[20] COULON, C., MIYASAKA, H., AND CLÉRAC, R. Single-Chain Magnets:Theoretical Approach and Experimental Systems. In *Single-Molecule Magnets and Related Phenomena*, R. Winpenny, Ed., no. 122 in Structure and Bonding. Springer Berlin Heidelberg, 2006, pp. 163–206.

[21] COURTENS, E. Vogel-Fulcher Scaling of the Susceptibility in a Mixed-Crystal Proton Glass. *Physical Review Letters 52*, 1 (Jan. 1984), 69–72.

[22] DALE, B. W., WILLIAMS, R. J. P., JOHNSON, C. E., AND THORP, T. L. S = 1 Spin State of Divalent Iron. I. Magnetic Properties of Phthalocyanine Iron (II). *The Journal of Chemical Physics 49*, 8 (Oct. 1968), 3441–3444.

[23] DAVIS, L. Oxidation Effects on the Magnetic Properties of Fe(II) Phthalocyanine Thin Films. Master's thesis, California State University Long Beach, Aug. 2017.

[24] DE DIESBACH, H., AND VON DER WEID, E. Quelques sels complexes des o-dinitriles avec le cuivre et la pyridine. *Helvetica Chimica Acta 10*, 1 (Jan. 1927), 886–888.

[25] DIXON, P. K., WU, L., NAGEL, S. R., WILLIAMS, B. D., AND CARINI, J. P. Scaling in the relaxation of supercooled liquids. *Physical Review Letters 65*, 9 (Aug. 1990), 1108–1111.

[26] EHRLICH, G. Atomic View of Surface Self-Diffusion: Tungsten on Tungsten. *The Journal of Chemical Physics 44* (1966), 1039.

[27] EKSTRAND, P. D., JAVIER, D. J., AND GREDIG, T. Tunable finite-sized chains to control magnetic relaxation. *Physical Review B 95*, 1 (Jan. 2017), 014406.

[28] ELEY, D. Phthalocyanines as Semiconductors : Abstract : Nature. *Nature 162* (Nov. 1948), 819.

[29] EVANGELISTI, M. *Experiments on Low-Dimensional Molecular Magents.* PhD thesis, University of Leiden, The Netherlands, 2001.

[30] EVANGELISTI, M., BARTOLOMÉ, J., DE JONGH, L. J., AND FILOTI, G. Magnetic properties of α-iron(II) phthalocyanine. *Physical Review B 66*, 14 (Oct. 2002), 144410.

[31] FERBINTEANU, M., MIYASAKA, H., WERNSDORFER, W., NAKATA, K., SUGIURA, K., YAMASHITA, M., COULON, C., AND CLÉRAC, R. Single-Chain Magnet (NEt$_4$)[Mn$_2$(5-MeOsalen)$_2$Fe(CN)$_6$] Made of Mn^{3+}-Fe^{3+}-Mn^{3+} Trinuclear Single-Molecule Magnet with an $S_T = 9/2$ Spin Ground State. *Journal of the American Chemical Society 127*, 9 (Mar. 2005), 3090–3099.

[32] FILOTI, G., KUZ'MIN, M. D., AND BARTOLOMÉ, J. Mössbauer study of the hyperfine interactions and spin dynamics in α-iron(II) phthalocyanine. *Physical Review B 74*, 13 (Oct. 2006), 134420.

[33] FORREST, S. R. The path to ubiquitous and low-cost organic electronic appliances on plastic. *Nature 428*, 6986 (Apr. 2004), 911–918.

[34] FRANK, P. *Thin Film Growth of Rod-like and Disc-Shaped Organic Molecules on Insulator and Noble Metal Surfaces.* Phd thesis, Graz University of Technology, July 2009.

[35] GENTRY, K. P., GREDIG, T., AND SCHULLER, I. K. Asymmetric grain distribution in phthalocyanine thin films. *Physical Review B 80*, 17 (Nov. 2009), 174118.

[36] GLAUBER, R. J. Time-Dependent Statistics of the Ising Model. *Journal of Mathematical Physics 4*, 2 (Feb. 1963), 294–307.

[37] GREDIG, T., COLESNIUC, C. N., CROOKER, S. A., AND SCHULLER, I. K. Substrate-controlled ferromagnetism in iron phthalocyanine films due to one-dimensional iron chains. *Physical Review B 86*, 1 (July 2012), 014409.

[38] GREDIG, T., GENTRY, K. P., COLESNIUC, C. N., AND SCHULLER, I. K. Control of magnetic properties in metallo-organic thin films. *Journal of Materials Science 45*, 18 (Sept. 2010), 5032–5035.

[39] GREDIG, T., SILVERSTEIN, E. A., AND BYRNE, M. P. Height-Height Correlation Function to Determine Grain Size in Iron Phthalocyanine Thin Films. *Journal of Physics: Conference Series 417*, 1 (Mar. 2013), 012069.

[40] GREDIG, T., WERBER, M., GUERRA, J. L., SILVERSTEIN, E. A., BYRNE, M. P., AND CACHA, B. G. Coercivity Control of Variable-Length Iron Chains in Phthalocyanine Thin Films. *Journal of Superconductivity and Novel Magnetism 25*, 7 (Oct. 2012), 2199–2203.

[41] GREGORY, P. *High-Technology Applications of Organic Colorants*, 1 edition ed. Springer, Aug. 1991.

[42] HLAWACEK, G., PUSCHNIG, P., FRANK, P., WINKLER, A., AMBROSCH-DRAXL, C., AND TEICHERT, C. Characterization of Step-Edge Barriers in Organic Thin-Film Growth. *Science 321*, 5885 (July 2008), 108–111.

[43] HU, Z., LI, B., ZHAO, A., YANG, J., AND HOU, J. G. Electronic and magnetic properties of metal phthalocyanines on Au (111) surface: A first-principles study. *The Journal of Physical Chemistry C 112*, 35 (2008), 13650–13655.

[44] ISING, E. Beitrag zur Theorie des Ferromagnetismus. *Zeitschrift für Physik 31*, 1 (Feb. 1925), 253–258.

[45] JIANG, J., Ed. *Functional Phthalocyanine Molecular Materials: 135*, 2010 edition ed. Springer, Feb. 2010.

[46] KANAMORI, J. Superexchange interaction and symmetry properties of electron orbitals. *Journal of Physics and Chemistry of Solids 10*, 2 (July 1959), 87–98.

[47] LAHERRÈRE, J., AND SORNETTE, D. Stretched exponential distributions in nature and economy: "fat tails" with characteristic scales. *The European Physical Journal B - Condensed Matter and Complex Systems 2*, 4 (Apr. 1998), 525–539.

[48] LEAL DA SILVA, J. K., MOREIRA, A. G., SILVÉRIO SOARES, M., AND SÁ BARRETO, F. C. Critical dynamics of the open Ising chain. *Physical Review E 52*, 4 (Oct. 1995), 4527–4528.

[49] LECREN, L., ROUBEAU, O., COULON, C., LI, Y.-G., LE GOFF, X. F., WERNSDORFER, W., MIYASAKA, H., AND CLÉRAC, R. Slow Relaxation in a One-Dimensional Rational Assembly of Antiferromagnetically Coupled [Mn$_4$] Single-Molecule Magnets. *Journal of the American Chemical Society 127*, 49 (Dec. 2005), 17353–17363.

[50] LEVER, A. B. P. The magnetic behaviour of transition-metal phathalocyanines. *Journal of the Chemical Society (Resumed)* (Jan. 1965), 1821–1829.

[51] LEVER, A. B. P., HEMPSTEAD, M. R., LEZNOFF, C. C., LIU, W., MELNIK, M., NEVIN, W. A., AND SEYMOUR, P. Recent studies in phthalocyanine chemistry. *Pure and Applied Chemistry 58*, 11 (2009), 1467–1476.

[52] LEZNOFF, C., AND LEVER, A. B. P. *Wiley: Phthalocyanines, Properties and Applications*, vol. 4. Wiley, May 1996.

[53] LIAO, M.-S., AND SCHEINER, S. Electronic structure and bonding in metal phthalocyanines, Metal=Fe, Co, Ni, Cu, Zn, Mg. *The Journal of Chemical Physics 114*, 22 (June 2001), 9780–9791.

[54] LINSTEAD, R. P. Phthalocyanines. Part I. A new type of synthetic colouring matters. *Journal of the Chemical Society (Resumed) 212*, 0 (Jan. 1934), 1016–1017.

[55] LINSTEAD, R. P., AND ROBERTSON, J. M. The stereochemistry of metallic phthalocyanines. *Journal of the Chemical Society (Resumed)* (Jan. 1936), 1736–1738.

[56] LUSCOMBE, J. H., LUBAN, M., AND REYNOLDS, J. P. Finite-size scaling of the Glauber model of critical dynamics. *Physical Review E 53*, 6 (June 1996), 5852–5860.

[57] MANRIQUEZ, J. M., YEE, G. T., MCLEAN, R. S., EP-
STEIN, A. J., AND MILLER, J. S. A Room-Temperature
Molecular/Organic-Based Magnet. *Science 252*, 5011 (June 1991),
1415–1417.

[58] MILLER, C. W., SHARONI, A., LIU, G., COLESNIUC, C. N.,
FRUHBERGER, B., AND SCHULLER, I. K. Quantitative structural
analysis of organic thin films: An x-ray diffraction study. *Physical
Review B 72*, 10 (Sept. 2005), 104113.

[59] MILLER, J. S. Organic Magnets—A History. *Advanced Materials
14*, 16 (Aug. 2002), 1105–1110.

[60] MILLER, J. S. Oliver Kahn Lecture: Composition and structure of
the V[TCNE]$_x$ (TCNE = tetracyanoethylene) room-temperature,
organic-based magnet – A personal perspective. *Polyhedron 28*,
9–10 (June 2009), 1596–1605.

[61] MILLER, J. S., AND EPSTEIN, A. J. Organic and Organometallic
Molecular Magnetic Materials—Designer Magnets. *Angewandte
Chemie International Edition in English 33*, 4 (1994), 385–415.

[62] MILLER, N. A. Role of Oxygen and Water Absorption on Charge
Transport in Copper Phthalocyanine Thin Films. Master's thesis,
California State University Long Beach, Aug. 2016.

[63] MUCKLEY, E. S., MILLER, N., GREDIG, T., AND IVANOV, I. N.
Effect of film morphology on oxygen and water interaction with
copper phthalocyanine. In *Proc. SPIE 9944, Organic Sensors and
Bioelectronics IX* (2016), pp. 99440V–9.

[64] MYDOSH, J. A. Spin glasses: Redux: An updated experimen-
tal/materials survey. *Reports on Progress in Physics 78*, 5 (2015),
052501.

[65] NATH, A., KOPELEV, N., TYAGI, S. D., CHECHERSKY, V., AND
WEI, Y. A novel class of ferromagnetic materials. *Materials Letters 16*, 1 (Feb. 1993), 39–44.

[66] NAUMIS, G. G., AND PHILLIPS, J. C. Bifurcation of stretched exponential relaxation in microscopically homogeneous glasses. *Journal of Non-Crystalline Solids 358*, 5 (Mar. 2012), 893–897.

[67] NEWELL, G. F., AND MONTROLL, E. W. On the Theory of the Ising Model of Ferromagnetism. *Reviews of Modern Physics 25*, 2 (Apr. 1953), 353–389.

[68] OGUCHI, T. Exchange Interactions in $Cu(NH_3)_4SO_4H_2O$. *Physical Review 133*, 4A (Feb. 1964), A1098–A1099.

[69] PALMER, R. G., STEIN, D. L., ABRAHAMS, E., AND ANDERSON, P. W. Models of Hierarchically Constrained Dynamics for Glassy Relaxation. *Physical Review Letters 53*, 10 (Sept. 1984), 958–961.

[70] PEIERLS, R. On Ising's model of ferromagnetism. *Mathematical Proceedings of the Cambridge Philosophical Society 32*, 3 (Oct. 1936), 477–481.

[71] PEISERT, H., SCHWIEGER, T., AUERHAMMER, J. M., KNUPFER, M., GOLDEN, M. S., FINK, J., BRESSLER, P. R., AND MAST, M. Order on disorder: Copper phthalocyanine thin films on technical substrates. *Journal of Applied Physics 90*, 1 (July 2001), 466–469.

[72] PETERSEN, J. L., SCHRAMM, C. S., STOJAKOVIC, D. R., HOFFMAN, B. M., AND MARKS, T. J. A new class of highly conductive molecular solids: The partially oxidized phthalocyanines. *Journal of the American Chemical Society 99*, 1 (Jan. 1977), 286–288.

[73] PHILLIPS, J. C. Stretched exponential relaxation in molecular and electronic glasses. *Reports on Progress in Physics 59*, 9 (1996), 1133.

[74] PHILLIPS, J. C. Microscopic aspects of Stretched Exponential Relaxation (SER) in homogeneous molecular and network glasses and polymers. *Journal of Non-Crystalline Solids 357*, 22–23 (Nov. 2011), 3853–3865.

[75] PINI, M. G., RETTORI, A., BOGANI, L., LASCIALFARI, A., MARIANI, M., CANESCHI, A., AND SESSOLI, R. Finite-size effects on the dynamic susceptibility of CoPhOMe single-chain molecular magnets in presence of a static magnetic field. *Physical Review B 84*, 9 (Sept. 2011), 094444.

[76] REIF, F. *Fundamentals of Statistical and Thermal Physics*. Mc-Graw Hill Higher Education, Auckland, Jan. 1965.

[77] ROSIN, P., AND RAMMLER, E. The Laws Governing the Fineness of Powdered Coal. *J. Inst. Fuel. 7* (1933), 29–36.

[78] RUIZ, R., CHOUDHARY, D., NICKEL, B., TOCCOLI, T., CHANG, K.-C., MAYER, A. C., CLANCY, P., BLAKELY, J. M., HEADRICK, R. L., IANNOTTA, S., AND MALLIARAS, G. G. Pentacene Thin Film Growth. *Chem. Mater. 16*, 23 (2004), 4497–4508.

[79] SANGREGORIO, C., OHM, T., PAULSEN, C., SESSOLI, R., AND GATTESCHI, D. Quantum Tunneling of the Magnetization in an Iron Cluster Nanomagnet. *Physical Review Letters 78*, 24 (June 1997), 4645–4648.

[80] SCHWOEBEL, R. L., AND SHIPSEY, E. J. Step Motion on Crystal Surfaces. *Journal of Applied Physics 37*, 10 (Sept. 1966), 3682–3686.

[81] SHEATS, J. R. Manufacturing and commercialization issues in organic electronics. *Journal of Materials Research 19*, 07 (2004), 1974–1989.

[82] SHTRIKMAN, S., AND WOHLFARTH, E. P. The theory of the Vogel-Fulcher law of spin glasses. *Physics Letters A 85*, 8 (Oct. 1981), 467–470.

[83] SKOMSKI, R., AND COEY, J. M. D. *Permanent Magnetism*. CRC Press, Bristol, UK ; Philadelphia, PA, Jan. 1999.

[84] STEER, C. A., BLUNDELL, S. J., COLDEA, A. I., MARSHALL, I. M., LANCASTER, T., BATTLE, P. D., GALLON, D., FARGUS, A. J., AND ROSSEINSKY, M. J. A μSR study of the spin dynamics in Ir-diluted layered manganites. *Physica B: Condensed Matter 326*, 1–4 (Feb. 2003), 513–517.

[85] VENABLES, J. A., SPILLER, G. D. T., AND HANBUCKEN, M. Nucleation and growth of thin films. *Reports on Progress in Physics 47*, 4 (1984), 399.

[86] VINDIGNI, A., BOGANI, L., GATTESCHI, D., SESSOLI, R., RETTORI, A., AND NOVAK, M. A. Finite size effects on the experimental observables of the Glauber model: A theoretical and experimental investigation. *Journal of Magnetism and Magnetic Materials 272–276, Part 1* (May 2004), 297–298.

[87] VINDIGNI, A., AND PINI, M. G. Selection rules for single-chain-magnet behaviour in non-collinear Ising systems. *Journal of Physics: Condensed Matter 21*, 23 (2009), 236007.

[88] VINDIGNI, A., RETTORI, A., PINI, M. G., CARBONE, C., AND GAMBARDELLA, P. Finite-sized Heisenberg chains and magnetism of one-dimensional metal systems. *Applied Physics A 82*, 3 (Oct. 2005), 385–394.

[89] WAGNER, H. J., LOUTFY, R. O., AND HSIAO, C.-K. Purification and characterization of phthalocyanines. *Journal of Materials Science 17*, 10 (Oct. 1982), 2781–2791.

[90] WANG, J., SHI, Y., CAO, J., AND WU, R. Magnetization and magnetic anisotropy of metallophthalocyanine molecules from the first principles calculations. *Applied Physics Letters 94*, 12 (Mar. 2009), 122502.

[91] WANG, Y., WU, K., KRÖGER, J., AND BERNDT, R. Review Article: Structures of phthalocyanine molecules on surfaces studied by STM. *AIP Advances 2*, 4 (Dec. 2012), 041402.

[92] WEI, R.-M., CAO, F., LI, J., YANG, L., HAN, Y., ZHANG, X.-L., ZHANG, Z., WANG, X.-Y., AND SONG, Y. Single-Chain Magnets Based on Octacyanotungstate with the Highest Energy Barriers for Cyanide Compounds. *Scientific Reports 6* (Apr. 2016).

[93] WERBER, M. Tuning Coercivity via Iron Chains in Phthalocyanine Thin Films. Master's thesis, California State University Long Beach, Long Beach, May 2013.

[94] WITTE, G., AND WÖLL, C. Growth of aromatic molecules on solid substrates for applications in organic electronics. *Journal of Materials Research 19*, 07 (Mar. 2011), 1889–1916.

[95] WÖHRLE, D., AND MEISSNER, D. Organic Solar Cells. *Advanced Materials 3*, 3 (Mar. 1991), 129–138.

[96] YANG, R. D., GREDIG, T., COLESNIUC, C. N., PARK, J., SCHULLER, I. K., TROGLER, W. C., AND KUMMEL, A. C. Ultrathin organic transistors for chemical sensing. *Applied Physics Letters 90*, 26 (June 2007), 263506.

[97] YOON, S. W., HEU, M., JEON, W. S., JUNG, D.-Y., SUH, B. J., AND YOON, S. Quantum tunneling and magnetic relaxation in Mn_{12} chloropropionate. *Physical Review B 67*, 5 (Feb. 2003), 052402.

[98] ZHANG, W.-X., ISHIKAWA, R., BREEDLOVE, B., AND YAMASHITA, M. Single-chain magnets: Beyond the Glauber model. *RSC Advances 3*, 12 (Feb. 2013), 3772–3798.

Index